Freeze-Fracture Replication

Freeze-Fracture Replication of Biological Tissues

Techniques, Interpretation and Applications

C. STOLINSKI
M.PHIL.
Lecturer in Biophysics, St. Mary's Hospital Medical School, University of London

A. S. BREATHNACH
M.SC., M.D.
Professor of Anatomy in the University of London at St. Mary's Hospital Medical School

1975

ACADEMIC PRESS
London New York San Francisco
A Subsidary of Harcourt Brace Jovanovich, Publishers

ACADEMIC PRESS INC. (LONDON) LTD
24/28 Oval Road, London NW1

United States Edition published by
ACADEMIC PRESS INC.
111 Fifth Avenue, New York, New York 10003

Library of Congress Catalog Card Number: 75–19678
ISBN: 0–12–672050–9

Printed in Great Britain by
Cox & Wyman Ltd,
London, Fakenham and Reading

Preface

This book has grown directly from the wish of each of us to apply a new technique to the study of a particular tissue, viz., blood and skin. Having seen the original Bullivant-Weinstein freeze-fracture apparatus at the Massachusetts General Hospital, we decided to design and build a module of our own, and we are grateful to Dr. Weinstein for his advice at the initial stages of this project. It was developed in co-operation with our colleagues, Maurice Gross and Barry Martin, and we are most appreciative of their invaluable assistance and continuing collaboration. Machining and perfecting of various items of equipment was done by the staff of the Department of Bio-medical Engineering of the Medical School, and generous financial support in the form of grants to the Department of Anatomy was provided by the Wellcome Trust, the Medical Research Council, the Peel Trust, the Fitton Trust, and the Joint Standing Research Committee of St. Mary's Hospital.

The general significance of the freeze-fracture replication technique, and the way in which it arose out of previous ultrastructural technical developments is outlined in Chapter 1. It will suffice here to state that it is now recognised as capable of making a unique contribution to the investigation of the functional morphology of biological tissues, particularly in relation to problems in membrane research and that it therefore merits the attention of all biologists. The increasing number of freeze-fracture papers appearing in the journals is a token of this. It is a somewhat complicated technique, presenting problems both in tissue processing and in the interpretation of the final electron micrographs. Whereas some aspects of these problems are dealt with in various rather specialised publications already available, we felt there was a need for a more basic general text written from the point of view of providing practical assistance to those interested in setting up facilities and applying the technique to their own particular fields of investigation. Having approached the matter ourselves from the ground level, we feel reasonably well qualified to draw attention to difficulties which may beset the uninitiated, and this book is the repository of our experience to date. The electron micrographs represent an important element of the book, and we are most grateful to Academic Press for allowing us to include so many, and to Mrs Georgina Klär for her efforts in ensuring a high standard of photographic reproduction. Finally, we should like to thank Miss Mary Lanyon for her impeccable typing of the manuscript.

Chapters 1–4 include some work contained in a thesis submitted by one of us (C.S.) in fulfilment of the requirements for award of the degree of M.Phil. of the University of London.

C. Stolinski
A. S. Breathnach

December, 1974

Contents

ISRAELIS CONRADI
Med. Doct.
DISSERTATIO
MEDICO-PHYSICA
DE
FRIGORIS
NATURA ET EFFE-
CTIBUS.

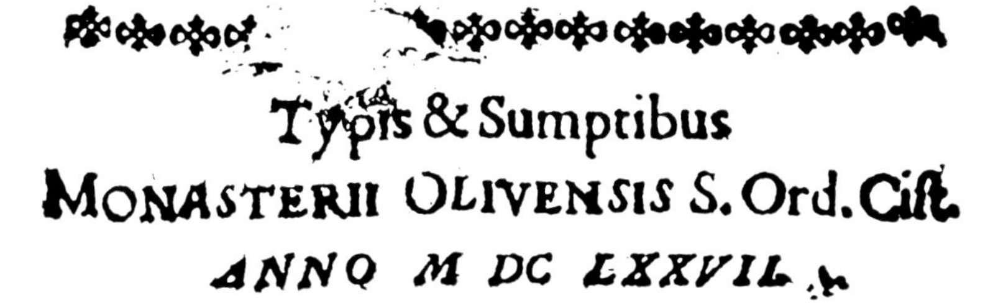
Typis & Sumptibus
MONASTERII OLIVENSIS S. Ord. Cist.
ANNO M DC LXXVII.

FIG. 1b. Title page of Conradt's (1677) book on the nature and effects of cold. Reproduced by permission of the British Library Board.

and carried out some experiments on the effect of severe cold on some animals (Fig. 1). He noticed that lampreys, gudgeons and frogs survived for short periods, even when ice had formed on their outer tissues, but concluded that prolonged cold could be a killer. Réaumur (1736) froze caterpillars using his newly designed thermometer to measure the cold, and observed species differences in survival after thawing, which he related to differences in the temperatures at which the blood froze.

One of the most colourful of the early scientists interested in the effect of cold, was the Abbate Lazzaro Spalanzani, first Professor of Natural History in the University of Pavia, and founder of its museum, where certain parts of his anatomy are still on view. Spallanzani was a great controversialist, having arguments with, amongst others, Buffon, Scarpa and Fontana. However, in between altercations and extensive travels, he found the time to conduct well-planned and precisely observed experiments on a variety of subjects (Fig. 2). Using rotifers and other 'animalcules', he observed the results of prolonged cooling, freezing, and supercooling. Animals which perished through being frozen solid at 6° 'below zero', were able to survive a temperature of 9° below zero in a fluid medium. Like Réaumur, Spallanzani (1776) observed species differences in survival rate, and he also noted that cold was less noxious to embryos ('germs') or eggs, than to adult animals. Insect eggs kept at a temperature of 17° below zero throughout the winter produced a normal new generation in the spring and this he attributed to the effect of 'oily fluids' within them. This observation foreshadowed the application of glycerol as a cryo-protective agent a century and a half later (Polge *et al.*, 1947).

The name of John Hunter inevitably crops up whatever aspect of the history of biology is being reviewed, and it is no surprise to find that in a very real sense he can be regarded as the great-great-great-grand-father of freeze-fracture (Stolinski, 1975). During the winter of 1777 he performed experiments to find out what truth there was in the current popular belief that animals could survive complete freezing. He was able to produce low temperatures in the vicinity of 0° (approximately −16 °C) by mixing ice and sea salt. A barrel filled with the cold mixture was then used to find the temperature at which blood would freeze. It was recorded at 25° (approximately −4°C), and Hunter concluded that animal tissues should also freeze near that temperature. The apparatus he devised for investigating this possibility bears such a striking resemblance to some used in modern freeze-fracture techniques that a quotation is justified:

> In January 1777, I mixed salt and ice till the cold was about 0° [−16°C]; On the side of the vessel was a hole, through which I introduced the ear of a rabbit. To carry the heat as fast as possible, it was held between two flat pieces of iron that went further into the mixture. That part of the ear projecting into the vessel became stiff, and when cut did not bleed; the part cut off by a pair of scissors flew from between the blades like a hard chip. (Hunter, 1779; p. 34).

Many other small animals and parts of live creatures were experimented on by Hunter and examined in the apparatus described. Freshly

OPUSCOLI
DI FISICA ANIMALE, E VEGETABILE

DELL'

ABATE SPALLANZANI

REGIO PROFESSORE DI STORIA NATURALE
NELL' UNIVERSITA' DI PAVIA;

SOCIO DELLE ACCADEMIE DI LONDRA, DE' CURIOSI DELLA NATURA DI GERMANIA, DI BERLINO, STOCKOLM, GOTTINGA, BOLOGNA, SIENA, ec.

Aggiuntevi alcune Lettere relative ad essi Opuscoli dal celebre Signor Bonnet di Ginevra, e da altri scritte all' Autore.

VOLUME PRIMO.

IN MODENA MDCCLXXVI.

PRESSO LA SOCIETA' TIPOGRAFICA.
Con Licenza de' Superiori.

FIG. 2. Title page of Spallanzani's work which includes his well-observed experiments on freezing of small animals. Photograph kindly supplied by Professor G. Moretti.

dissected muscles of lambs, combs and wattels of young cocks, tails of gold fishes, various eggs, tree saplings and seeds, were subjected to freezing. Despite the fact that he was more interested in the revival of his experimental objects after thawing, the detailed examination of fractured faces of animal tissues fixed by freezing, had occurred to him as being of interest. He was without a doubt the first individual to freeze-fracture and investigate animal tissues, and the rabbit's ears referred to above are still on exhibition at the Hunterian Museum of the Royal College of Surgeons of England (Fig. 3).

At the beginning of the nineteenth century, Rudolphi, when studying filaria, decided to freeze them and to test their survivability. He remarked that 'a worm must hold firmly to life when, after being preserved frozen rigid and covered with hard ice for 8 days, on melting, the filaria comes to life again' (Rudolphi, 1808). Pouchet (1866) made an early attempt to freeze and preserve blood, but he noticed that on thawing the blood became 'soluble', i.e. haemolysis had set in. By the end of the nineteenth century, following construction of the first successful air liquifier, biologists started to freeze a variety of samples down to -200 °C. Brown and Escombe (1897) tested survivability of bacteria after prolonged immersion in liquid air, as well as seeds for their germinative power. In Germany, Altman (1890) recognised the potentialities of freezing as an alternative method for the fixation of tissues. He hit on the idea of sublimating frozen tissues *in vacuo* and subsequently examining the dry specimens at room temperature. However, the drawbacks of the technique, caused mainly by the growth of ice crystals damaging the tissues during freezing, were recognised by Molisch (1897) and the technique was abandoned. It was revived much later by Gersch (1932) and applied eventually as a very successful industrial and laboratory freeze-drying technique. With this technique, freezing is used successfully as a preliminary step. However, the subsequent complete ice sublimation and substitution results in redistribution of structures and the possible production of specific artifacts (Bernhard and Leduc, 1967).

At the beginning of the twentieth century many authors published work connected with the experimental freezing of living forms. Plank *et al.* (1916), recognised the importance of the rate at which the samples were cooled, and Chambers and Hale (1932) studied the formation of ice crystals in protoplasm. Luyet and Gehenio (1940) produced a

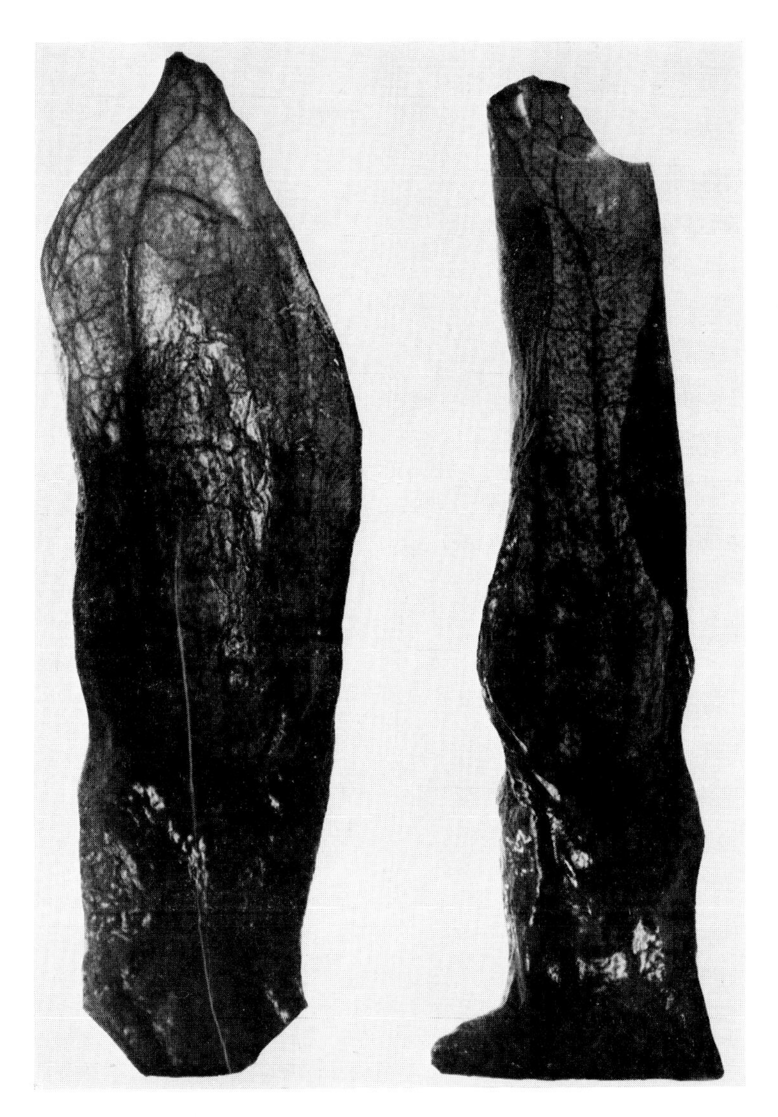

FIG. 3. Rabbit's ear frozen and fractured by John Hunter in the year 1777. On the left is a control ear and on the right, the frozen ear from which a portion of the tip was fractured away. Reproduced by permission of the President and Council of the Royal College of Surgeons of England.

monograph in which a summary of their researches was published with the main conclusion that it is essential to produce a vitreous state of ice in the tissues in order for them to survive solidification. Later on, successful experiments involving freezing and revival of vinegar eels and frog spermatozoa were performed after cooling them in liquid air to −190 °C.

II MODERN STUDIES, AND DEVELOPMENT OF FREEZE-FRACTURE TECHNIQUES

Great impetus was added to research on the freezing of living material by the discovery of the protective action of glycerol by Polge *et al.* in 1947. Mammalian and avian sperms, as well as cells and tissues, were protected successfully against the harmful effects of freezing and thawing. It then became possible to store samples in the frozen state and in the presence of glycerol without loss of viability at −190 °C for long periods of time. In subsequent years it has also been discovered that protection with glycerol against the damage produced by freezing is not a completely artificial means devised by man. In nature, the blood of some hardy insects has up to 20% free glycerol, which helps them to survive through the most severe winters (Salt, 1958). Parkes (1957), with the co-operation of the Royal Society, organised a meeting in London at which the viability of mammalian tissues and cells subjected to freezing was discussed. Almost every aspect of the field was reviewed by an international gathering of eminent workers. Among many topics discussed were the practical application of freezing to transplantation of tissues, transfusion of blood, and storage of sperms for fertilisation. A. U. Smith later produced excellent reviews covering the field up to 1960 (Smith, 1961). More recently P. Mazur, a notable researcher in the field of cryobiology, has written a comprehensive review covering a variety of aspects of work connected with freezing of biological systems (Mazur, 1970). This work provided an essential background for the successful development of the freeze-fracture technique.

Advances achieved with techniques of electron microscopy in the 1950s, led some workers to realise that methods in vogue for tissue preparation were unsatisfactory. Chemical fixation, complete dehydration, embedding and staining, were almost certainly introducing artifacts, and it was difficult to establish just how true a picture of actual structure was being portrayed. Freezing had been tried for some time as a method of tissue fixation, but with only moderate results (Sjöstrand, 1943). Following extensive experiments, Wyckoff (1946, 1949) and his team (Williams and Wyckoff, 1944) developed a technique of freeze-drying combined with metal-shadowing, which, when applied to a variety of important biological molecules, produced micrographs of high quality; excellent surface replicas of skin were also produced. This pioneer work, which can be regarded as the starting-point of the freeze-fracture replication technique was followed by Hall's (1950) and Meryman's (1950) independent construction of apparatus for replication of frozen samples with silicon monoxide. Hall (1950), who investigated ice and other crystals, was also able to utilise and control sublimation of ice to provide added relief. Subsequently, Meryman and Kafig (1955) applied the technique to biological material, with little success. Steere (1957) modified the ideas of Hall and Meryman, and further developed the technique to the point which is today regarded as its first successful application to biological tissues. He froze crystallised T.M.V. viruses and cracked them in a special cold box. The sample was then transferred to a vacuum evaporator where replicas were produced (Fig. 4). Haggis (1961) later successfully produced replicas of cracked red cells and for the first time the platinum-carbon shadowing technique of Bradley (1958) was employed. By breaking inside the vacuum chamber a thin glass cover-slip covered with frozen cells, Haggis claimed that the haemoglobin molecules became visible. However, membranes were not demonstrated and no further progress was made. It was finally Moor *et al.* (1961) who, returning to the ideas of Hall, Meryman and Steere, fully realised the potentialities of freeze-fracture replication.

The technique which they called 'freeze-etching', involved chipping frozen samples with a precision, liquid nitrogen-cooled, ultramicrotome placed inside a vacuum evaporator chamber. They produced a great variety of good replicas of glycerinated and quickly frozen biological materials (Moor, 1964), and a completely new picture of biological ultrastructure emerged. A three-dimensional topography was revealed consisting of splits and fractures, and in particular, large areas of biological membranes (Fig. 5). A commercial apparatus (Balzer's) based upon Moor's design was subsequently put on the market (Moor,

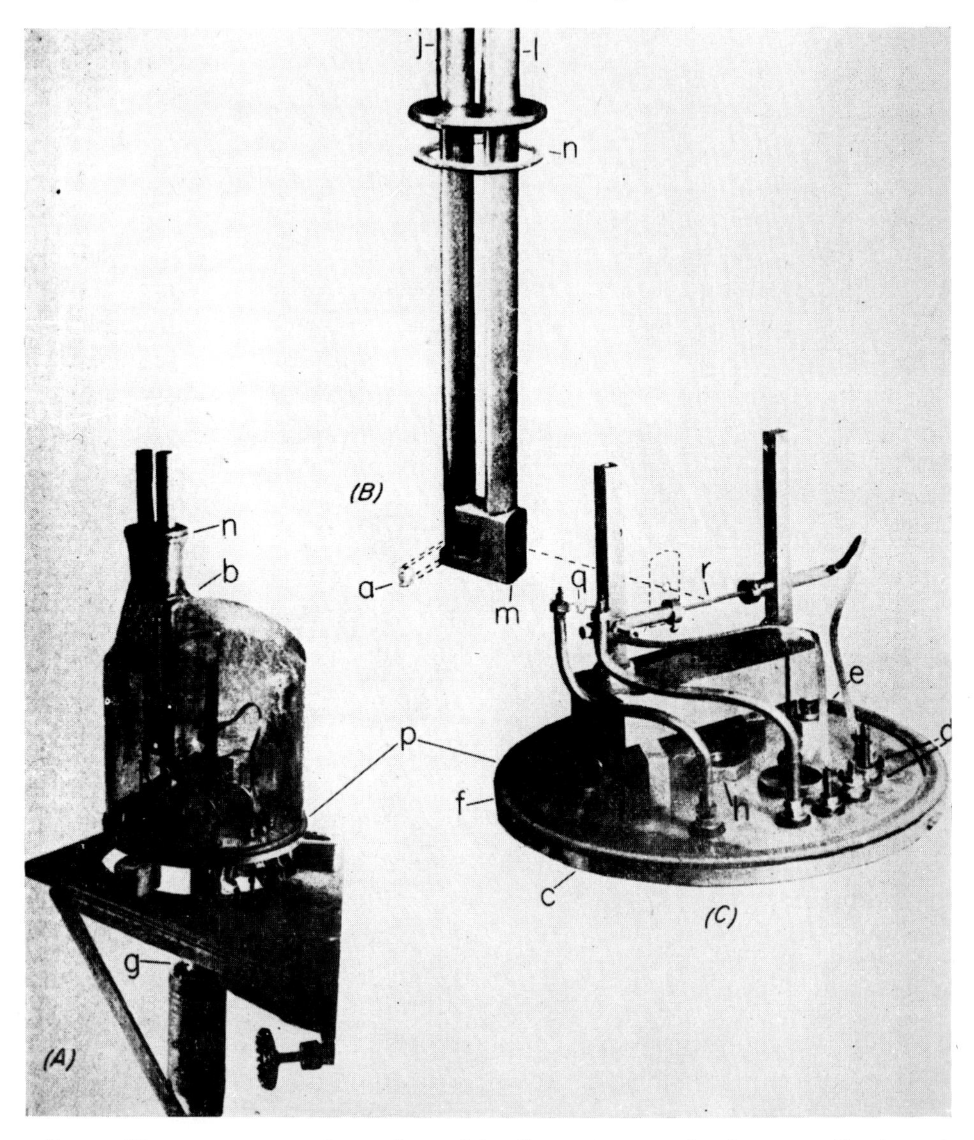

FIG. 4. Vacuum evaporation unit used by Steere in 1957 for freeze-replication of virus crystals. The excellent quality of Steere's pictures clearly demonstrated the potentialities of the technique. Reproduced by permission of Dr. Steere, the United States Department of Agriculture, and Rockefeller University Press.

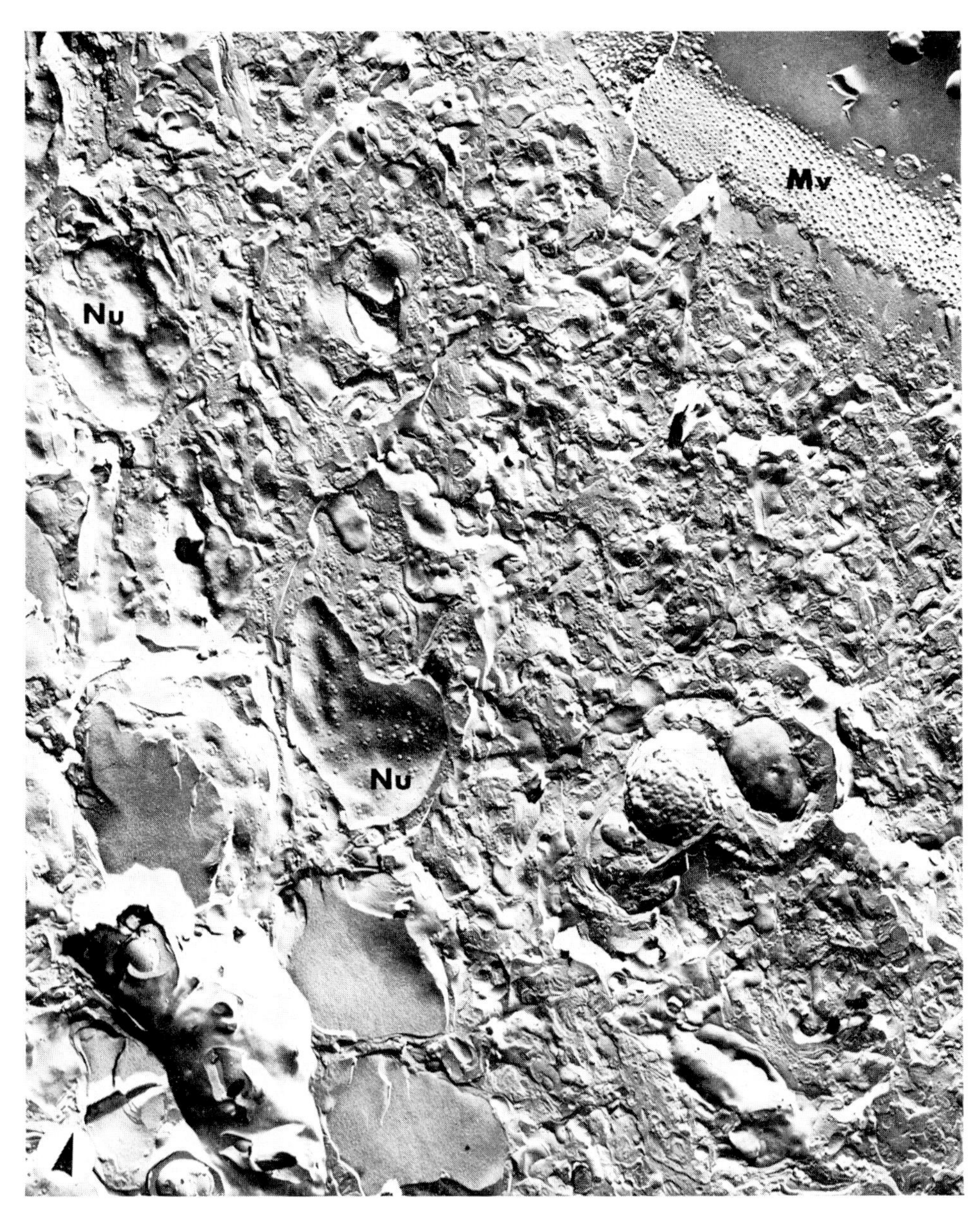

Fig. 5. Freeze-fracture replica of mouse jejeunal epithelium. A three-dimensional topography is apparent in these shadowed replicas. Nu, nucleus; Mv, microvilli of brush border. × 6000. Arrow at bottom left indicates direction of shadowing.

1965), and this prompted several investigators to build their own versions of a freeze-cracker. Bullivant and Ames (1966) questioned the necessity of using an ultramicrotome to produce fractures. Their samples, after freezing, were fractured under liquid nitrogen, transferred to a vacuum evaporator, still under liquid nitrogen, which was then pumped out revealing the fractured tissue which was then shadowed and replicated. Essentially the same types of replicas as those produced by Moor's apparatus were produced. Others produced versions of Moor's and Bullivant's method where the sample was fractured inside (Steere, 1966; McAlear and Kreutziger, 1967; Boeve and Kreger, 1968) or outside (Geymeyer, 1966) the vacuum chamber. Recent developments in instrumentation have been concerned with eliminating water-vapour and oil contamination of the specimen while it is in the vacuum chamber, with more precise control and maintaining of temperature for sublimation, with securing better vacua, with producing matched replicas, and with the application of electron and ion guns for evaporation of shadowing materials. At the same time improved methods of specimen preparation are being investigated with the aim of discovering new and better cryoprotective agents, or of eliminating altogether the necessity for their application by means of ultra-rapid freezing.

The earliest studies employing freeze-fracture replication in biological research were carried out on isolated plant or animal cells such as yeast, fungi, algae, bacteria, platelets and erythrocytes, but soon botanists were examining more organised tissues such as root tips, etc. As regards animal tissues, the attention of pioneer investigators was directed towards individual components, or particular features such as myelin, muscle filaments, photoreceptors, and the plasma membrane and its specialised contacts. Recently, however, organised tissues such as liver, skin, brain, etc., have been examined more extensively from the point of view of establishing overall basic normal appearances, and comparing them with appearances as revealed by electron microscopy of thin sections. Some studies on pathological material have also been published. At regular intervals, excellent critical reviews have appeared discussing the general merits, significance, limitations, and applications of the technique, assessing progress to date, and pointing the way towards future developments (Moor, 1966, 1969, 1971; Weinstein and Someda, 1967; Koehler, 1968; Branton, 1971; Benedetti and Favard,

1973; Yago *et al.*, 1974). It is clear from these that the technique has unique advantages as a method for studying the functional morphology of biological tissues at the ultrastructural level, and that it provides a valuable supplement to already established techniques of light and electron microscopy.

REFERENCES

Altmann, R. (1890). 'Die Elementar-organismen und ihre Beziehungen zur den Zellen.' Viet, Leipzig.

Benedetti, E. L. and Favard, P. (eds) (1973). 'Freeze-etching, Techniques and Applications.' Société Française de Microscopie Électronique, Paris.

Bernhard, W. and Leduc, E. H. (1967). *J. Cell Biol.* **34**, 757–786.

Boeve, H. and Kreger, D. R. (1968). *Science Tools* **16**, 27–30.

Boyle, R. (1683). 'New Experiments and Observations Touching Cold.' Richard Davis, London.

Bradley, D. E. (1958). *Nature* **181**, 875–877.

Branton, D. (1971). *Phil. Trans. Roy. Soc.* **B261**, 133–138.

Brown, H. T. and Escombe, F. (1897). *Phil. Trans. Roy. Soc.* **62,** 160–173.

Bullivant, S. and Ames, A. (1966). *J. Cell. Biol.* **29**, 435–447.

Chambers, R. and Hale, H. D. (1932). *Phil. Trans. Roy. Soc.* **B110**, 336–352.

Clark, A. (ed.) (1898). 'Brief Lives, Chiefly of Contemporaries, Set Down by John Aubrey Between the Years 1609 and 1696.' Clarendon Press, Oxford.

Conradt, I. (1677). 'Dissertatio Medico-physica de Frigoris Natura et Effectibus.' Typis et Sumptibus Monastarii Olivensis, Gdansk.

Gersch, I. (1932). *Anat Rec.* **53**, 309–337.

Geymeyer, W. (1966). *Proc. 6th Int. Cong. E. M. Kyoto* **1**, 577–578.

Haggis, J. H. (1961), *J. Biophys. Biochem. Cytol.* **9**, 841–852.

Hall, C. E. (1950). *J. Appl. Phys.* **21**, 61–62.

Hunter, J. (1779). *Phil. Trans. Roy. Soc.* **68**, 7–49.

Koehler, J. K. (1968). *Adv. Biol. Med. Phys.* **12**, 1–84.

Luyet, B. J. and Gehenio, P. (1940). 'Life and Death at Low Temperatures.' Biodynamica Monograph. Normandy, U.S.A.

McAlear, J. H. and Kreutziger, G. O. (1967). *Proc. 25th Ann. E.M.S.A.*, B–34.

Mazur, P. (1970). *Science* **168**, 939–949.

Meryman, H. T. (1950). *J. Appl. Phys.* **21**, 68.

Meryman, H. T. and Kafig, E. (1955). *Nav. med. res. inst. project*, NM.000 018 01. **13**, 529–544.

Molisch, H. (1897). 'Untersuchungen über das Erfrieren der Pflanzen.' Fischer, Jena.

Moor, H. (1964). *Z. Zellforsch.* **62**, 546–580.

Moor, H. (1965). 'Balzer's High Vacuum Report.' No. 2, Liechtenstein.
Moor, H. (1966). *Int. Rev. Exp. Path.* **5**, 179–216.
Moor, H. (1969). *Int. Rev. Cytol.* **25**, 391–412.
Moor, H. (1971). *Phil. Trans. Roy. Soc.* **B261**, 121–131.
Moor, H., Mühlethaler, K., Waldner, H. and Frey-Wyssling, H. (1961). *J. Biophys. Biochem. Cytol.* **10**, 1–13.
Parkes, A. S. (1957). *Proc. Roy. Soc.* **B**147, 423–553.
Plank, R. Ehrenbaum, E. and Reuter, K. (1916). 'Die Konservierung von Fischen durch das Gefrierverfahren.' Teil II, Zentral Einkaufsgesellschaft, Berlin.
Polge, C., Smith, A. U. and Parkes, A. S. (1947). *Nature* **164**, 666.
Pouchet, P. (1866). *J. Anat. (Paris)* **3**, 1–36.
Power, H. (1663). 'Experimental Philosophy in Three Books Containing New Experiments.' Quoted by D. Keilin (1959), *Proc. Roy. Soc.* **B150**, 149–191.
Réaumur, R. A. F. de (1736). 'Memoires pour Servir à l'Histoire des Insectes.' Imprimerie Royale, Paris.
Rudolphi, C. A. (1808). 'Entozoorum Sire Vermium Intestinalium Historia Naturalis.' Vol II. Brockhaus, Leipzig.
Salt, R. W. (1958). *Nature* **184**, 1281.
Sjöstrand, F. S. (1943). *Nature* **151**, 725–726.
Smith, A. U. (1961). 'Biological Effects of Freezing and Supercooling.' Arnold, London.
Spallanzani, L. (1776). 'Opusculi di Fisica Animale e Vegetabile.' Presso la Societa Tipographica, Modena.
Steere, R. L. (1957). *J. Biophys. Biochem. Cytol.* **3**, 45–60.
Steere, R. L. (1966). *J. Appl. Phys.* **37**, 3939.
Stolinski, C. (1975). *Med. Hist.* **19**, 303–306.
Weinstein, R. S. and Someda, K. (1967). *Cryobiology* **4**, 116–129.
Williams, R. C. and Wyckoff, R. W. G. (1944). *J. Appl. Phys.* **21**, 61–62.
Wyckoff, R. W. G. (1946). *Science* **104**, 36–37.
Wyckoff, R. W. G. (1949). 'Electron Microscopy. Technique and Applications.' Interscience Publishers, Inc. New York.
Yago, J. H., Piqueras, J. R. and Bover, G. F. (1974). 'Criofractura y Ultra-structura Celular.' Editorial Facta, Valencia.

2

Theory and Principles of the Technique

I GENERAL OUTLINE OF PROCEDURE

In the following introductory paragraph the freeze-fracture replication technique is briefly outlined so as to provide a general background to the subsequent discussion of the underlying principles.

In order to avoid the damaging effects of freezing, the fresh sample of tissue is permeated with a cryoprotective or anti-freeze agent. In the majority of cases glycerol (up to 25% by volume) is used, and introduced into the sample at reduced temperatures. Alternatively, the sample may be chemically fixed and subsequently permeated with a cryoprotectant of choice. Small portions (2 to 3 cubic mm) of the pre-treated sample are then rapidly frozen in cold fluids, such as liquid propane, to a temperature in the vicinity of −180 °C. The solid frozen specimen on a metal support is then transferred into the vacuum chamber of an evaporator, where it is mounted on a cooled stage. Under vacuum, the specimen is cleaved or fractured, preferably near −120°C, so that a fresh previously unexposed surface is revealed (Fig. 6A). After fracture, sublimation of ice from the exposed fracture surface (etching) may or may not be carried out (Fig. 6B). Next, platinum-carbon (or other shadowing material) is evaporated at an angle on to the specimen. Platinum-carbon electron-dense material produces characteristic shadows which accentuate the local topography, while carbon evaporated normally completes the process by binding the isolated patches of evaporated platinum into a complete replica (Fig. 6C). The specimen with the replica is then removed from the vacuum chamber and thawed. The tissue substrate is dissolved or digested away, and the replica is cleaned and mounted on a supporting grid (Fig. 6D). It is then ready for examination in the electron microscope and eventual photography.

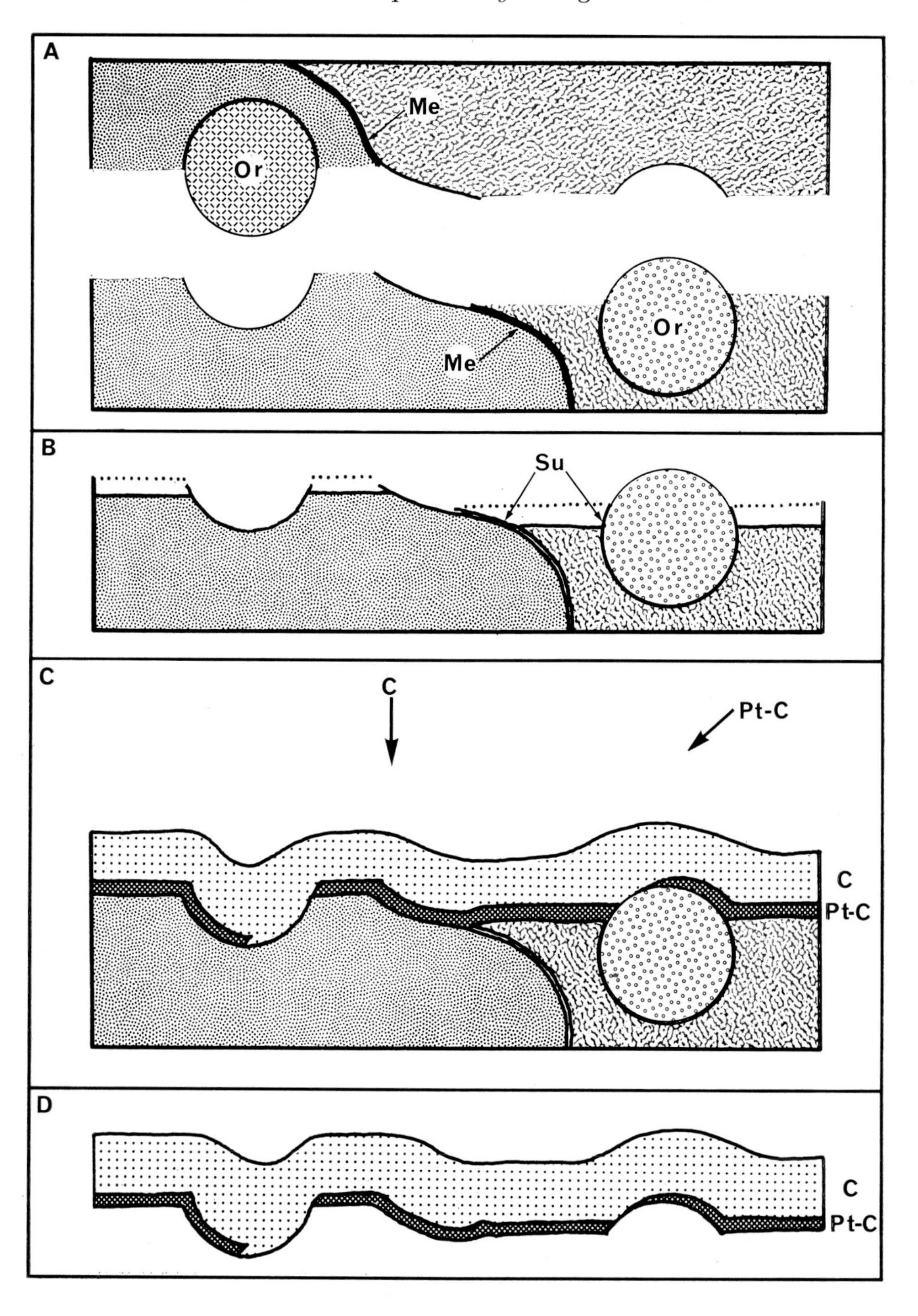
A
Me
Or
Me
Or
B
Su
C
C
Pt-C
C
Pt-C
D
C
Pt-C

II FREEZING AS A METHOD OF FIXATION

The prime feasability of the technique revolves around a unique method of physical fixation of tissues, namely, freezing. Water in its liquid state, within a narrow range of temperatures, is a necessary prerequisite of life. Frozen tissues, with an original amount of water retained in a solid state, can be considered as entities whose structure is preserved and whose life processes have been stopped, though remaining dormant. Carefully executed freezing does not destroy tissue viability, as shown in long-term storage of cells for fertilisation, transplantation and transfusion. Noticeable physiological damage is not evident in these frozen tissues, as most intricate metabolic activities can be revived on thawing and put into use on demand (Smith, 1961; Turner, 1970). Some experiments described by Andjus *et al.* (1955) increase confidence in freezing as a method of tissue preservation even further. They found that trained adult rats which were frozen and then thawed, retained, after revival, memory of the layout of a maze with hidden food. In general, the condition of frozen organisms where metabolic activities are dormant can be described in the context of microscopical preparatory techniques as 'fixed', and one would expect them to be unaltered structurally. However, some methods of freezing for the preservation of tissues tend to dehydrate the cells, resulting in shrinkage. The cells, which on thawing, may still be capable of being revived, are not completely satisfactory in their frozen state for morphological examination as some of their components may be distorted. Nevertheless, it is evident that this method of fixation, as compared to the chemical one, does not greatly alter the tissues, which must remain very close to the living state. The potential advantages of 'biological solid state as a unique and useful research tool', have been very aptly pointed out by

FIG. 6. Schematic representation illustrating in cross-sectional profiles the freeze-fracture replication procedure which consists of: A, fracturing of frozen tissue fragment, where two organelles (Or) are shown separated by a membrane, Me. B, if required, subsequent sublimation of the material surrounding the tissue components with the resulting lowering of the original level (shown with a dotted line) and exposure of some additional material, Su. C, formation of the replica on the fractured surface by first evaporating platinum-carbon (Pt–C) at an angle to form shadows, and then carbon (C) normally to the surface. (Pt–C and C layers are not to scale.) D, separation of replica from the underlying tissue.

Meryman (1956) in his review on the mechanics of freezing in biological systems. The freeze-fracture replication technique in association with electron microscopy visualises tissue ultrastructure in this unique state.

A Phenomena associated with Freezing

In order to be able to follow the complex factors involved in freezing of tissues, it is necessary to consider first of all what happens when water is frozen. When the temperature is lowered, ice does not form at 0 °C, but the water, still in a liquid state, is super-cooled to a temperature well below its freezing point (Fig. 7). At this stage, ice crystals begin to

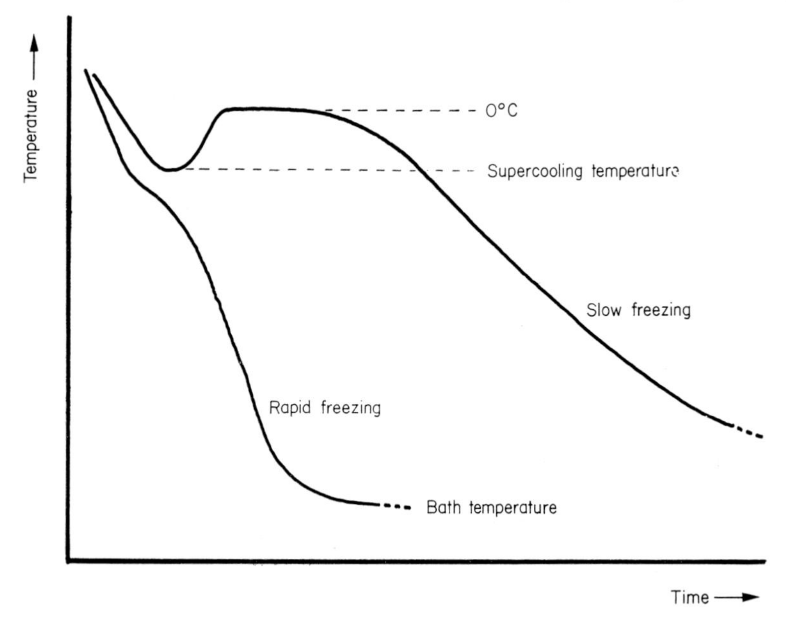

FIG. 7. Rate of change of temperature of small quantity of water when frozen slowly and rapidly.

form around impurities, or, in the case of very pure water, around specific clusters of water molecules. Heat, known as 'heat of crystallisation', is released at this stage and raises the temperature of the water up to its melting point, so that a mixture of ice and water results. At slow rates of freezing the temperature of this mixture can be further lowered only when all the water is changed to ice, at which stage the now solid material can be cooled to whatever desired temperature. However, if, during this entire process, heat of crystallisation can be efficiently extracted, by producing a sufficiently rapid rate of cooling, it is found that the temperature does not rise, and the water becomes super-cooled and frozen in a so-called 'vitreous' state. This consists of a solid state made up of numerous but very small ice crystals. Efficient extraction of heat of crystallisation by very rapid cooling, in effect, prevents the formation of large ice crystals.

1 Freezing of tissues and resulting damage. Since most biological tissues contain more than 80% water, the above considerations must be borne in mind if freezing is to be used as a preparative or preservative technique. Freezing, *per se,* performed without due caution, causes irreversible damage to living tissues (Lovelock, 1953; Meryman, 1956; Smith, 1961; Mazur, 1965, 1970). This damage can be categorised as follows:

1. Mechanical damage, caused by the growth of ice crystals inside or outside the cells which puncture membranes and distort organelles.
2. Dehydration damage, resulting from the increased concentration of salts in tissue fluids following crystallisation of water. This leads to the formation of harmful osmotic forces which can lead to excessive and harmful shrinkage or other damage.
3. 'Thermal shock', defined as damage caused by stresses induced by uneven expansion or contraction of cell components. Within the thermal shock injury, an effect described by Levitt and Dear (1970) can be included. This involves free proteins, which, during dehydration and freezing, may come into contact and form a permanent new structure through linkage of di-sulphide bonds.

All the above categories of specific damage can be assessed by physiological tests performed on previously frozen and thawed tissue. Farrant (1965), for example, tested the functional activity of smooth muscle after it was thawed from −79 °C. The results when compared with those derived from a non-frozen control, allowed Farrant to assess the damage and to point to the most effective method for freezing of tissue. The damage can be generally described in such cases as physiological damage. It should be remembered, however, that the degree of injury to the tissue as measured by such methods, includes the damage suffered by it during the thawing process. It follows, therefore, that with the freeze-fracture technique the overall morphological damage is less, as thawing is not involved in the process.

In order to prevent the above types of damage, and maintain tissues in a structurally unchanged form for subsequent examination, ways and means must be found to protect them during freezing, i.e. to allow them to become vitrified as completely and as soon as possible. This involves, essentially, methods for rapid freezing, and the application of cryoprotective or anti-freeze agents, usually in combination.

B Protection of Tissues from Freezing Damage

1 Rapid freezing of samples without cryoprotectant. Following some original ideas on freezing of sprayed blood (Meryman and Kafig, 1955) and on sprayed samples for freeze-drying (Williams, 1953), Bachman and Schmidt (1971) devised a method of freezing samples for freeze-fracture replication which shows considerable promise. The sample in the form of a suspension is sprayed into a cooled container filled with an appropriate cooling agent, such as liquid propane at −190 °C. Due to the small size of the droplets produced by the sprayer – 10 to 20 μm in diameter – a sufficiently rapid freezing rate of 50,000 °C per second is achieved so as to ensure effective vitrification of the sample. This technique, however, is applicable only to solutions and suspensions of cells which can be sprayed (Plattner *et al.*, 1973). It is difficult to see how it could be applied to organised multicellular tissue.

Another method of freezing tissues where cooling rates of the order of 100,000 °C per second can be achieved has been tried by Riehle and Hoechli (1973). The method involved subjecting the tissue to 2,000 bar pressure during cooling, and complete vitrification of samples was claimed. However, high-pressure rapid-freezing of this type introduces its own specific artifacts and difficulties (Moor, 1971), and has not yet been sufficiently widely performed to allow it to be realistically evaluated.

In actual practice, vitrification can be achieved at much lower rates of cooling (600 °C to 1,000 °C per second) if tissue samples are permeated by a cryoprotective agent prior to freezing.

2 Rapid freezing of samples in the presence of cryoprotectant. Liquid nitrogen is a very suitable intermediate cooling liquid, but if it is employed to freeze samples of standard useful size (2 cubic mm) it must be used in combination with another agent. If samples are immersed in liquid nitrogen alone, the rate of cooling is considerably slowed down by a layer of gaseous nitrogen which boils when it comes in contact with the warmer sample. The insulating effect produced by this layer of nitrogen gas (Leidenfrost's phenomenon) results in an inadequate cooling rate of 50°C per second. If liquid nitrogen is cooled to the vicinity of its melting point (−210°C) it is found that Leidenfrost's phenomenon does not occur, (Umrath, 1974) and rapid freezing rates are readily achieved. However special apparatus is necessary to produce the required low temperature. More conventional and traditional methods for rapid freezing involve primary cooling fluids like Freon, Isopentane or liquid Propane. Provided the boiling point of the cooling liquid is in the vicinity of the temperature of the sample, the specimen cooling rate following immersion will be in the region of over 600°C per second. The exchange of heat in this case is enhanced because the cold fluid remains in direct contact with the specimen until the bath temperature is reached by the tissue. Once the required low temperature is achieved, it can be maintained by use of liquid nitrogen.

3 Cryoprotective agents. These are agents used to permeate tissues in order to protect them from the damaging effects of freezing at intermediate (600° to 1000°C per second) rates of cooling. As water becomes bound to the molecules of the agent, a substantial lowering of the freezing point of tissue fluids results, so that the time available for the growth of ice crystals is shortened: smaller and therefore less destructive crystals are formed (Lussena, 1955). Lowering of the freezing point tends to shift the occurrence of damaging electrolyte concentration to temperatures at which cell components may be less soluble. Also, reduction of the relative concentration of electrolytes prevents the dehydration and osmotic damage already referred to.

The cryoprotective agent most commonly used for freeze-fracture is glycerol (Moore, 1964). Another agent, dimethyl sulphoxide (DMSO) also offers good protection (Lovelock and Bishop, 1959). It has been used extensively for the preservation of organs and tissues for transplantation. A great variety of other agents of low and high molecular weights have also been tried with reasonable success for other purposes (Turner, 1970; Meryman, 1971), but to date, only glycerol and DMSO have been used for the freeze-fracture technique.

It is quite likely that permeation of tissues with a cryoprotective agent can result in some damage or alteration, though precise details are not known. It can be argued that application of cryoprotective agents prior to freezing, alters appreciably the original water content of the specimen. However, such agents as a rule are non-toxic, the tissues can live and survive in their presence fairly well, and finally, the agents can be washed away without harming the sample. At any rate, at the present moment, their advantages can be said greatly to outweigh any disadvantages apparent. Ideally, in practice, one should aim at employing the lowest concentration of cryoprotectant consistent with obtaining a suitably rapid freezing rate and no formation of ice crystals. The phenomena associated with freezing of glycerol solutions at different rates and concentrations have been studied by Lussena (1955), and by Staehelin and Bertaud (1971). It is difficult however, to draw conclusions from these researches as to what should be the optimum concentration of glycerol in tissues. It probably varies with the tissue, and with whether or not it has been chemically fixed. Much further experimentation is required in this connection, and details of current practice are provided in the following chapter.

Freezing even at optimum rate and in the presence of correct amount of cryoprotective agent, can still alter morphology (Meryman, 1956). For example, some molecules that constitute protein gels can be dissolved by cooling (Marsland, 1956). Another characteristic example illustrating the solubility of semicrystalline protein aggregates which liquify on lowering the temperature can be observed in deoxygenated sickle haemoglobin (Murayama, 1965).

III FRACTURING OF FROZEN SAMPLES

With the freeze-fracture technique, the interior of the frozen tissue is exposed by inducing or guiding a fracture through it, and a variety of methods have been employed to achieve this. It may be cleaved in half outside the vacuum chamber with a cooled blade or other device (Steere, 1957; Bullivant and Ames, 1966). However, in order to be able to replicate immediately freshly exposed surfaces, the sample may be introduced intact into a vacuum chamber and chipped with a cold blade attached to a precision microtome (Moor *et al.*, 1961), or parted with a cleaver specially designed for that purpose (Stolinski, 1975).

Whatever method may be used, a fracture surface of the tissue is exposed, which is essentially uneven. The crack, or fissure, which is influenced by the forces exerted on the sample, follows the plane where least resistance is offered to it. It is very apparent from micrographs that the fissure generated within the tissue is greatly influenced by discontinuities associated with the membranes. As the fracture traverses a frozen cell, it traverses the cytoplasm in a random fashion, but as it approaches membranes – plasma, nuclear, or those surrounding organelles – it is frequently deviated, and runs parallel with the membrane for variable distances, thus exposing extensive areas *en face* (Fig. 8). (Membranes may also be fractured straight across, thus appearing as in thin sections.) It is this more extensive revelation of membranes which makes the technique so potentially valuable.

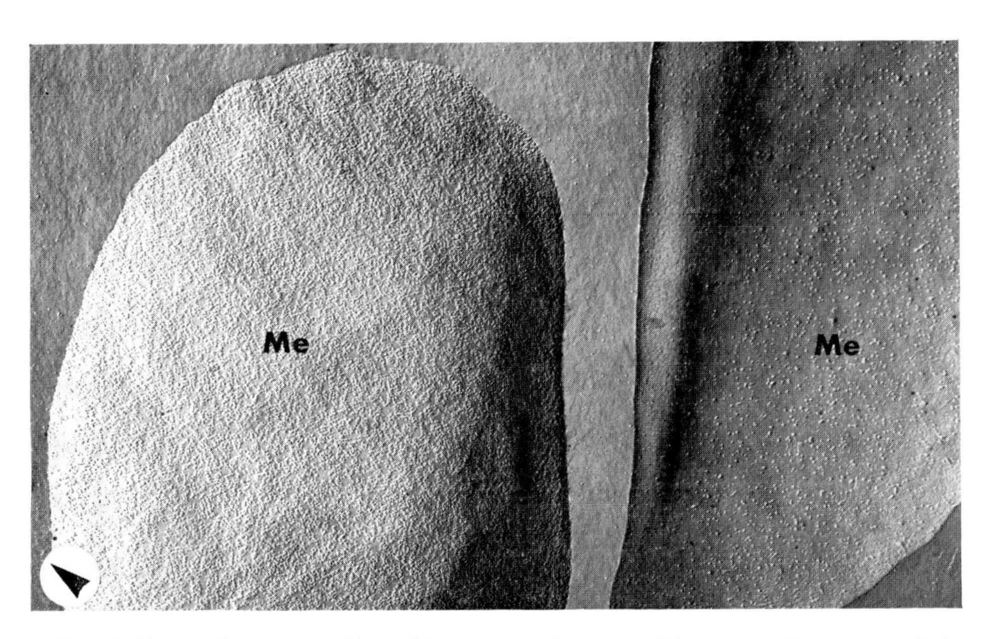

Fig. 8. Freeze-fracture replica of human erythrocytes. Note extensive areas which are closely associated with the plane of the membrane, Me. × 26,000.

A Matched Fractures

In most instances, only one of the two surfaces resulting from fracture of a specimen is retained for replication. However, for certain purposes, and in order to provide additional information relating mainly to membrane structure, it is desirable to replicate both surfaces. This allows one to compare exactly matching areas from each side of the fracture, which theoretically should be the reflex one of the other. Production of matched replicas presents formidable technical problems in the design of specimen holders, finder grids, etc., but these have been successfully surmounted in different ways as described in the following chapter.

B *Sublimation of Ice (Etching)*

Following fracture, the surface of the specimen which is exposed, will consist of formed tissue elements embedded in ice, or in a frozen glycerol-water mixture. This surface may be immediately replicated (freeze-fracture replication), or it may be partly freeze-dried before replication by raising the temperature to the vicinity of −100°C for a minute or so. This is the step commonly referred to as 'etching' or 'freeze-etching', and it involves a sublimation process whereby ice molecules are converted directly into vapour which disperses in the vacuum. The result of this removal of ice is a lowering of the general level of the fracture surface, and the revelation of additional areas of non-sublimatable biological material which were beneath the original ice table (Figs. 6 and 9). The rate and extent to which sublimation can occur depends upon a number of factors (Moor, 1973), one of which is temperature, and another of which is the concentration of cryoprotective agent employed. No effective sublimation occurs from the surface of tissue permeated with 30% glycerol, though such material is often presented in the literature as having been 'freeze-etched', mainly because it was processed in a commercial 'freeze-etching' machine. The term 'etching' to describe the process of sublimation is increasingly being regarded as unsatisfactory, but it is fairly well entrenched in the literature already, and all one can do is to advocate its use as little as possible in the hope that it will be gradually phased out.

The possibility of freeze-drying and exposing specimens surrounded by ice was first realised by Wyckoff (1946). Later, controlled sublimation was elaborated and analysed by Hall (1950). Subsequently Davy and Branton (1970) described the sublimation process in detail, explaining the appearance of asperites formed on affected surfaces.

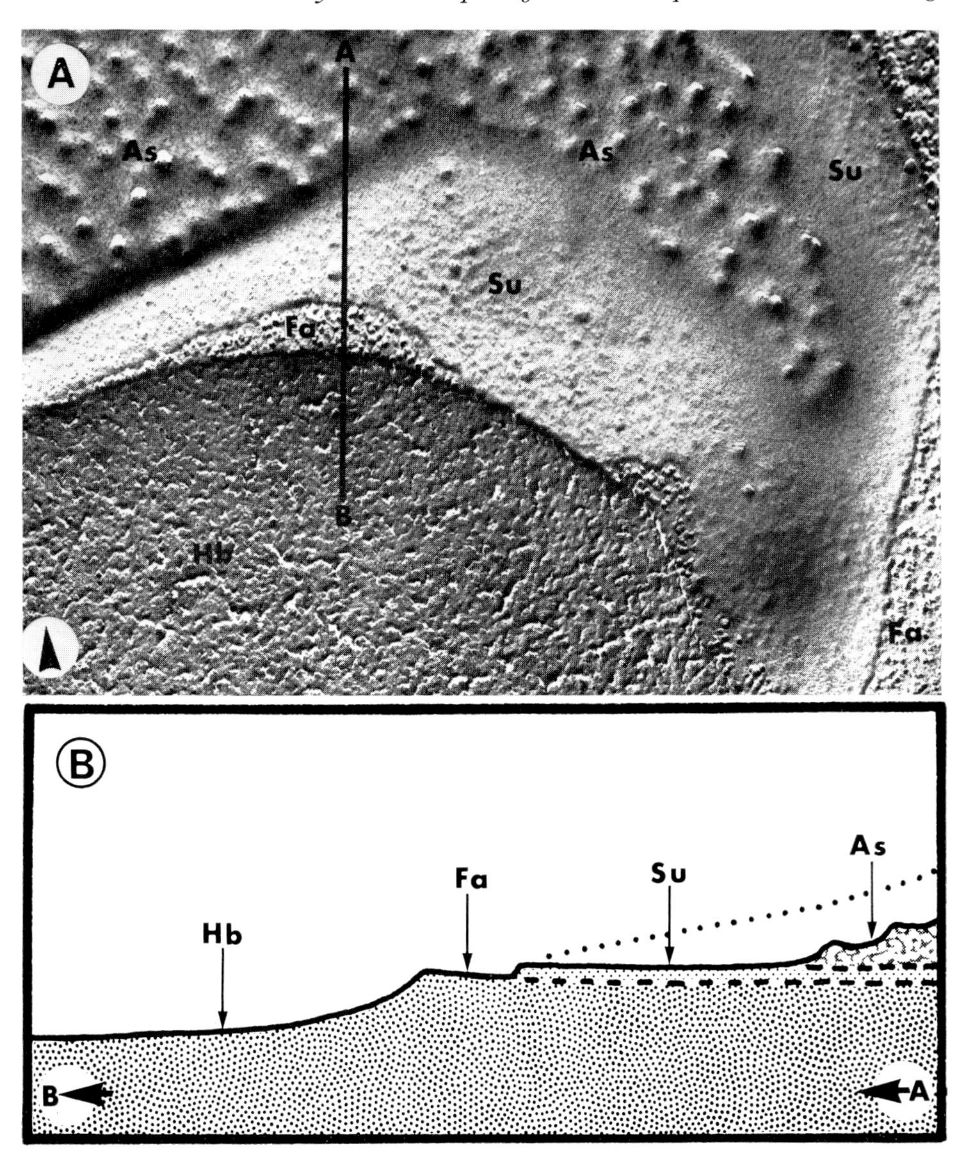

FIG. 9. A. Replica of human erythrocyte showing areas of material (Su) revealed by the sublimation process. The material thus revealed is situated on top of the fracture face (Fa) previously exposed by fracturing. As, asperites formed during sublimation process; Hb, haemoglobin. × 86,500. B. Profile of fracture faces along the line A–B. Dotted line represents the level of the original surface of the ice.

Deep sublimation of small or unicellular specimens suspended and frozen in aqueous media has illuminated some fundamental aspects of the technique in general, as will be explained in Chapter 4, but it has been applied only with limited success to larger samples of organised tissues. This is due mainly to the non-sublimatable character of tissue cryoprotected by glycerol, but the more general application of other cryoprotectives such as DMSO, which will allow effective sublimation of reasonably sized samples of non-chemically fixed tissue, should soon be feasible. However, it must be borne in mind that vacuum sublimation is a somewhat violent process capable of producing artifacts which can complicate interpretation (Moor, 1971, 1973; Staehelin and Bertaud, 1971; Lickfeld *et al.*, 1972). As a general rule, in the investigation of any tissue the uncomplicated freeze-fracture appearances should first be established before attempts at sublimation are embarked upon.

IV REPLICATION

The aim of replication is the production of a recoverable facsimile of the fracture surface, on which the topographical relief has been accentuated by shadowing. The most commonly employed technique derives from Bradley's (1954, 1958) method for evaporation of platinum and carbon from a low voltage arc. Moor's (1959) modification of the method made a practical contribution by incorporating a spiral of thin platinum wire in the carbon electrodes. Shadows are produced by evaporating platinum-carbon at an angle to the plane of the fracture, and carbon alone is then evaporated normal to the specimen. The low electron-dense carbon effectively coats the whole surface of the specimen and binds all the isolated particles of platinum carbon into a replica (Fig. 10). Recently, newly developed methods of evaporation with an electron gun are beginning to be adopted (Bachmann *et al.*, 1969; Zingsheim *et al.*, 1970; Moor, 1973), and in our laboratory the application of ion guns is being investigated, so far with limited success. It is hoped that these newer methods may produce better resolution and that the evaporation process may be rendered simpler and more controllable than it is with conventional arc technique.

Recovery of a clean intact replica presents a number of difficulties. After the specimen is removed from the vacuum, the tissue to which the

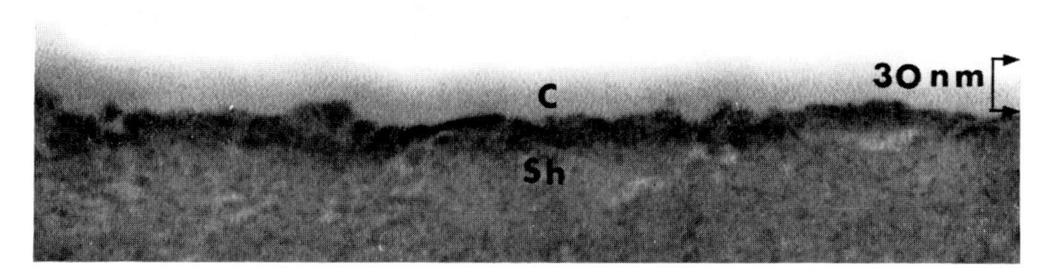

FIG. 10. Curved portion of a replica observed tangentially. A substantial thickness of evaporated carbon film, C, is observed on top of the shadowing material, Sh. × 220,000.

replica is bound has to be thawed, and then separated or digested away from the replica by various reagents. During these processes, swelling or shrinkage of the tissue can occur leading to damage and fragmentation of the inelastic replica. The degree of difficulty encountered at this stage depends to a great extent on the tissue involved; skin, for example, is particularly onerous to deal with. Each worker tends to have his own favourite method, and a limited amount of general guidance exists in the literature; our particular methods are outlined in the next chapter.

V ARTIFACTS

As with any other method of processing biological tissues for observation, possible sources of artifact which can affect interpretation must be considered, and, if possible, eliminated. Apart from tissue damage due to freezing referred to already, artifacts can arise (or be introduced) at other stages of the technique. Consideration of these, from the point of view of recognition and avoidance, is best deferred to the following chapter dealing with practical details of the steps at which they may arise.

VI NOMENCLATURE

Precise definition and general acceptance of descriptive terms is essential for the intelligible communication of knowledge arising out of the application of any new technique. With freeze-fracture, whereas workers in general are gradually moving towards a standardised terminology, there is still some diversity of usage in the literature, and this can be confusing for the uninitiated. Until, at some international congress or other, a generally agreed terminology is promulgated, individual workers can but ensure that whatever terms they use, they should be clearly defined and readily understood. In practice we have found the following terms suitable for defining and explaining our observations and interpretations and relating them to those of other workers:

Fracture: the dynamic process which induces a discontinuity (or crack) in the specimen, and which results in its separation into two parts.

Fracture plane: plane of discontinuity in the solid material induced by external forces.

Fracture face: either of the two faces which are revealed by fracturing a specimen.

Rough or random fracture face: fracture face associated with a non-organised composite material consisting of various molecules and ice; for example, cytoplasmic material.

Fissure face: fracture face closely associated with, and parallel to a biological membrane.

Fracture face type 'Ext': fissure face associated with a membrane and directed towards the material that is external to the substance enclosed by that membrane. Can apply not only to plasma membrane, but also to membranes limiting intracellular organelles, e.g. mitochondrial membranes. This type of face is referred to by many authors as the 'A-face' of the membrane (Fig. 11).

Fracture face type 'Int': fissure face associated with a membrane and directed towards the material that is internal to, or enclosed by that membrane. Can apply equally to plasma membranes and membranes limiting cytoplasmic organelles. This type of face is referred to by many authors as the 'B-face' of the membrane (Fig. 11). It is suggested that the terms 'Ext' and 'Int' are preferable to 'A' and 'B' (and to others which have been used, such as 'convex' and 'concave', or 'positive' and 'negative'), since they give some indication of the nature and disposition of the fracture faces to which they refer.

Split membrane face: a general term to describe a fracture face revealed by a fissure produced in the interior of a membrane. Originally used by authors who specified a fracture plane coinciding with the exact centre of the membrane, it is now applied to any membrane fracture face revealed by interior fission whatever the spatial position, and therefore embraces faces 'Ext' and 'Int' as defined above.

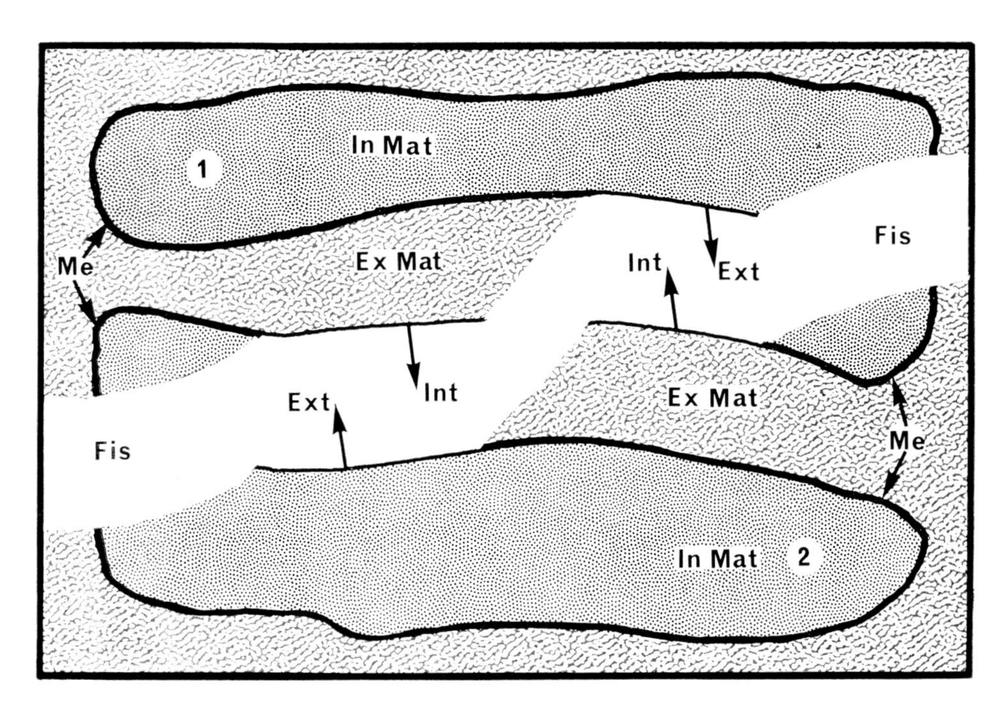

FIG. 11. Diagram to illustrate terminology used in describing fracture faces associated with biological membranes. Two cells (1, 2) surrounded by external material (Ex Mat) are shown separated by a fissure (Fis) produced during the fracturing process. Me, membrane enclosing internal material. Ext, fracture face directed towards external material; Int, fracture face directed towards internal material (In. Mat).

Matched fracture faces: fracture faces indentified on each part of the fractured specimen when both are replicated, and which were originally in direct contact or continuity.

Complementary fracture faces: fissure, or split-membrane faces revealed on replicas of one part (side) of a fractured specimen, and representing 'Ext' and 'Int' type faces of individually different membranes of similar type. Such faces are frequently observed at the apposition of two cells of the same type. An 'Ext' type face belonging to the plasma membrane of one cell is separated by intercellular material from an 'Int' type face associated with the plasma membrane of the second cell. These faces are said to be complementary, and corresponding matched faces for each would be present on the material fractured away. Theoretically, bringing together of complementary fracture faces should reconstitute a membrane, but not one that ever existed in fact, because the material involved would be derived from plasma membranes of two different cells. Where several similar organelles are present within a cell, complementary fracture faces of the limiting membrane – each from a different organelle are frequently seen on the same replica.

Sublimation or 'etching': A process of controlled vacuum freeze-drying involving sublimation of ice from the surface exposed by fracturing or any other surface. This can reveal additional areas of biological material which were beneath the original ice table.

REFERENCES

Andjus, R. H., Knopfelmacher, F., Russel, R. W. and Smith, A. U. (1955). *Nature, Lond.* **176**, 1015–1016.

Bachmann, L. and Schmidt, W. W. (1971). *Proc. Nat. Acad. Sci. U.S.A.* **68**, 2149–2152.

Bachmann, L., Abermann, R. and Zingsheim, H. P. (1969). *Histochemie* **20**, 133–142.

Bradley, D. E. (1954). *Brit. J. App. Phys.* **5**, 65–66.

Bradley, D. E. (1958). *Nature, Lond.* **181,** 875–877.

Bullivant, S. and Ames, A. (1966). *J. Cell Biol.* **29**, 435–447.

Davy, J. G. and Branton, D. (1970). *Science* **168**, 1216–1218.

Farrant, J. (1965). *Nature, Lond.* **205**, 1284–1287.

Hall, C. E. (1950). *J. Appl. Physics* **21**, 61–62.

Lickfeld, K. G., Achterrath, M. and Hentrick, F., (1972). *J. Ultrastruct. Res.* **38**, 279–287.

Levitt, J. and Dear, J. (1970). *In* 'The Frozen Cell' (G. E. W. Wolstenholme and Maeve O'Connor, eds) 149–174. Churchill, London.

Lovelock, J. E. (1953). *Biochem. Biophys. Acta* **11**, 28–36.

Lovelock, J. E. and Bishop, M. W. H. (1959). *Nature, Lond.* **183**, 1394–1395.

Lussena, C. V. (1955). *Arch. Biochem. Biophys.* **57**, 277–284.

Marsland, D. (1956). *Int. Rev. Cytol.* **5**, 199–226.

Mazur, P. (1965). *Fed. Proc. Suppl.* **15**, S175–S182.

Mazur, P. (1970). *Science* **168**, 939–949.

Meryman, H. T. (1956). *Science* **124**, 515–521.

Meryman, H. T. (1971). *Cryobiology* **8**, 173–183.
Meryman, H. T. and Kafig, E. (1955). *Proc. Soc. Exptl. Biol. Med.* **90**, 587–589.
Moor, H. (1959). *J. Ultrastruct. Res.* **2**, 393–422.
Moor, H. (1964). *Z. Zellforsch.* **62**, 546–580.
Moor, H. (1971). *Phil. Trans. Roy. Soc. Lond.* **B261**, 121–131.
Moor, H. (1973). *In* 'Freeze-Etching, techniques and applications' (E. L. Benedetti and P. Favard, eds) 21–26. Société Française de Microscopie Électronique, Paris.
Moor, H., Mühlethaler, K., Waldner, H. and Frey-Wyssling, A. (1961). *J. Biophys. Biochem. Cytol.* **10**, 1–13.
Murayama, M. (1966). *Science* **153**, 145–149.
Plattner, H., Schmitt-Fumian, W. W. and Bachmann, L. (1973). *In* 'Freeze-Etching, techniques and applications' (E. L. Benedetti and P. Favard, eds) 81–100. Société Française de Microscopie Électronique, Paris.
Riele, U. and Hoechli, M. (1973). *In* 'Freeze-Etching, techniques and applications' (E. L. Benedetti and P. Favard, eds) 31–61. Société Française de Microscopie Électronique, Paris.
Smith, A. U. (1961). 'Biological Effects of Freezing and Supercooling.' Arnold, London.
Staehelin, L. A. and Bertaud, W. S. (1971). *J. Ultrastruct. Res.* **37**, 146–168.
Steere, R. L. (1957). *J. Biophys. Biochem. Cytol.* **3**, 45–60.
Stolinski, C. (1975). *J. Micros.* **104** (in press).
Turner, A. R. (1970) 'Frozen Blood: A review of Literature 1949–1969.' Gordon and Breach, New York.
Umrath, W. (1974). *J. Micros.* **101**, 103–105.
Williams, R. C. (1953). *Exptl. Cell Res.* **4**, 188–201.
Wyckoff, R. W. G. (1946). *Science* **104**, 36–37.
Zingsheim. H. P., Abermann, R. and Bachmann, L. (1970). *J. Physics E: Sci. Instr.* **3**, 39–42.

3

Practical Aspects of Specimen Processing

Successful preparation of replicas from freeze-fractured material requires acquisition of skills in processing and manipulating specimens, as well as in the correct and efficient utilisation of apparatus. As regards the latter, a number of excellent custom-built machines are available commercially which are supplied complete with appropriate instructions for use. However, right from the start of the application of the technique, individual workers have designed their own freeze-fracture modules for attachment to standard vacuum-coating units of the type available in most electron microscope laboratories. This is a practice which will probably be continued for some time by the impecunious with access to facilities for making such modules. In this chapter, apart from practical considerations which have general application whatever the apparatus used, we also provide some guidelines from our own experience for those contemplating designing their own module.

I PERMEATION OF SPECIMENS WITH CRYOPROTECTIVE AGENTS

With spray-freezing methods (see p. 20), and with certain tissues of low water content, such as stratum corneum of skin, cryoprotection is not necessary; otherwise it is essential in order to avoid damage due to freezing.

A Chemical Pre-fixation of Samples

A review of the literature reveals that the great majority of workers investigating animal tissues routinely employ chemical fixation prior to the application of cryoprotective agents. This is done with the intention of reducing tissue necrosis, and of eliminating artifacts allegedly induced by cryoprotective agents (MacIntyre *et al.*, 1974). However, chemical fixation leads to modifications of fine detail in the final micrograph (Dempsey *et al.*, 1973) and it can modify the cleavage plane of membranes (James and Branton, 1971); it impedes the penetration of cryoprotective agents, besides which, its employment vitiates one of the main advantages of the technique, which is the possibility of investigation of tissues unaffected by the many various processes involved in preparing them for conventional transmission electron microscopy. Also, it is possible that the artifacts which MacIntyre *et al.* (1974) attributed to cryoprotectants, were in fact due to the low temperature (0°C) at which their material was maintained throughout. Speth *et al.* (1973), and Wunderlich *et al.* (1973, 1974), have described very similar low temperature effects. These considerations make it very difficult to know what advice should be given over this matter of chemical fixation. On balance, perhaps some degree of chemical artifact or loss of detail is preferable to partial necrosis. So, if in doubt, and particularly with scarce or unique material, or with tough specimens which will probably be permeated slowly by the cryoprotective, it would appear advisable to fix chemically. With readily available material, both fixed and unfixed specimens may be examined. Routinely adopted, at least for some time to come, this practice would help to clarify some issues at present in doubt concerning the nature and source of artifacts, and might, in addition, permit more precise definition of some morphological features. The chemical fixative most commonly used is buffered 1–3% glutaraldehyde, for periods up to 1 hour. All micrographs in this book are of unfixed material.

B Cryoprotection with Glycerol

Glycerol solutions of 20 to 30% concentration by volume are the most commonly employed cryoprotectants. They are made up by replacing appropriate volumes of water in buffered (pH 7·4) physiological solutions by glycerol. The solution we have found most satisfactory is a 25% solution in TC199 concentrate reached in two stages, with a first stage of 15%. Others have employed Cacodylate, or phosphate buffers, or, Ringer-Tyrode, in making up glycerol solutions.

Particularly with specimens which have not been pre-fixed chemically, rapid permeation with the cryoprotectant is essential. Permeation can be speeded up considerably if the tissue in its fluid container is placed in a rotating tumbler which prevents the build-up of stagnant layers of viscous glycerol around the tissue. The minimum time necessary for adequate permeation will depend upon the tissue, the size of the specimen, whether it was pre-fixed chemically or not, and on the temperature, and indeed there are practically no data available which could provide guidelines on these matters. Times of actual immersion in glycerol solutions as quoted in the literature vary between 20 minutes and overnight, and where specified, temperatures have usually been between 0° and 4°C. For example, our practice with unfixed samples of liver is to place 2 mm^3 blocks in the 15% glycerol solution for 20 minutes, and for a further 30 minutes in the 25% solution, employing a battery-operated tumbler which is placed in the refrigerator. Carrying out permeation at less than room temperature should help to reduce necrotic changes in unfixed specimens, but it can of itself produce certain alterations in membranes (Wunderlich *et al.*, 1974). So, once again, one is between the devil and the deep.

In the case of cell suspensions like blood, the sample should be mixed with an equal volume of buffered solution containing a proportion of cryoprotective agent in place of water. Prior to experiment, an appropriate amount of stock buffer solution can be mixed with an equal amount of 100 or 80% solution of the cryoprotective agent chosen, to form final solutions by volume of, respectively, 50 and 40% cryoprotective agent in a physiologically balanced buffer solution. Finally, the cells, suspended in their natural fluids, are mixed at room temperature for 20 minutes with the prepared solutions at equal volumes. The final concentration of 25% glycerol results in a reasonable cryoprotection. The concentration of cells in the final solution will also be adequate provided that the supernatant of the suspension was removed, prior to loading of the specimen holders.

With bacteria, it is often the practice to freeze non-cryoprotected samples, which were chemically fixed (Nanninga, 1973). No doubt in such instances some freezing damage is sustained by the sample, but effective sublimation can be performed. For adequate cryoprotection, glycerol up to 20% concentration by volume in a buffered solution, applied for 1 hour can be employed, but in such instances sublimation is not feasible. The application of DMSO, provided its toxity can be tolerated, and the use of spray freezing methods seem to be the methods of choice for bacteria. Application of DMSO should prevent freezing damage and it does not hinder sublimation.

Botanical samples are best permeated by being grown in a medium which is gradually enriched in the desired amount of cryoprotectant. For example, by growing onion root tips, for 4 to 7 days in a solution of 20% glycerol in water good protection is achieved (Branton and Moor, 1964).

C *Application of DMSO*

Application of DMSO involves procedures essentially similar to glycerination. Solutions, however, cannot be stored over long periods of time, and have to be prepared freshly for each application. So far, DMSO has been extensively tried only with cell suspensions (MacIntyre *et al.*, 1974), where permeation with 30 to 40% DMSO in buffered solution mixed for 20 minutes with an equal volume of sample was found to be adequate.

The main advantages of DMSO are that it reduces freezing damage while at the same time permitting sublimation (Stolinski, 1973). However, it is found to be toxic in elevated concentrations, and for that reason, the permeation period should be as brief as possible (Alink *et al.*, 1974).

II FREEZING OF SPECIMENS

After permeation of the sample with cryoprotectant is completed, it is introduced into or on to an appropriate holder, depending upon the apparatus used for fracturing. Some holders by virtue of their design, prevent possible loss of the entire sample on fracturing. These holders, which are in the form of small stainless steel wells or tubes, hold the bulk of the sample in their interior by virtue of firm adhesion to the side walls of the container, the small portion to be fractured away projecting beyond the upper edge. Other holders are in the form of discs appropriately scored to enhance contact between sample and metal.

A *Spray Freezing*

Freezing whole cells, or cellular components, suspended in a suitable medium, by the method of spray freezing, is one of the latest innovations in the technique and shows considerable promise for future development (Bachmann and Schmitt-Fumian, 1973). The supporting apparatus required to perform such freezing, is as yet not available commercially, but it can if necessary be constructed locally. For producing very fine droplets in the region of 10 μm diameter a commercial spray gun as used by draughtsmen can be successfully employed. Directing such a sprayer into a bath containing liquid propane, results in small droplets of suspension containing the specimen becoming so rapidly frozen that ice crystals are not formed (Bachmann and Schmitt, 1971). Subsequently, by slightly raising the temperature of the freezing fluid it can be evaporated away and substituted with a substance capable of binding the isolated droplets. *N*-butylbenzene fulfils this function as it is liquid at −80 °C and freezes solid at −88 °C. Droplets in the suspending medium at −80 °C can subsequently be loaded in cooled specimen holders maintained at the same temperature and frozen solid together with the binding medium, in liquid nitrogen. Fracturing and all subsequent stages in the production of replicas are then identical to routine operations.

B Routine Freezing of Cryoprotected Specimens

Melting nitrogen (liquid nitrogen cooled to −210°C) propane, Freon 22, and Isopentane (see Appendix for properties), are the most commonly used primary cooling fluids for rapid freezing. On the whole melting nitrogen is the most satisfactory (Umrath, 1974) as it can reach a temperature in the vicinity of −200 °C while still in the liquid state, whereas Isopentane and Freon 22 freeze solid at −160 °C, and Propane at −180 °C. Faster freezing rates, therefore, are feasible when melting nitrogen is used (Mackenzie, 1969).

Initial freezing fluid in a small container is cooled by liquid nitrogen and the specimen in its holder is plunged in. After the specimen is frozen it is withdrawn from the cold bath, slightly warmed up, and the excess of coolants like Propane or Isopentane which tend to cling to the holder are removed by touching it lightly with filter paper. This frozen fluid, if not removed will be carried into the vacuum chamber where it will sublimate and contaminate the freshly fractured specimen (see Fig. 21). Then, the holder with frozen specimen is transferred to a container of liquid nitrogen, beneath the surface of which it may be introduced into a relatively large metal holder which allows it to be transferred and firmly mounted on to the cold stage of the vacuum chamber. With some apparatus where intermediate holders are not employed, the transfer must be executed very quickly.

After rapid freezing, frozen specimens in their holders may be stored in a liquid nitrogen container for fracturing at some future date. Such specimens may be stored almost indefinitely without undergoing deterioration. If specimens are being stored, it is advisable again to remove the excess of initial coolant, before placing them in the storage container.

This completes the prime processing of the specimen, and its subsequent handling will depend to some extent upon the apparatus available. In what follows, therefore, only general practical considerations relating to the basic design and manner of working of freeze-fracture apparatus are dealt with.

C *Transfer of Specimen to Vacuum Chamber*

Transfer of the sample into the vacuum chamber and its firm mounting on to a pre-cooled stage should be smoothly and quickly executed. In this way excessive warming up as well as collection of large amounts of frost on the cold sample can be avoided. Venting of the vacuum chamber, should be done with dry nitrogen gas delivered from a gas cylinder at approximately 5 lb per sq. in, while gaining access to it to introduce the specimen. Even after access to the chamber has been gained the flow of gas should be maintained. This continuous flow effectively excludes atmospheric moisture, and prevents formation of frost on cold surfaces inside the chamber. If nitrogen gas is not available, air introduced into the chamber should be dried by passing it over silica gel.

III VACUUM CONDITIONS

After the sample on its support has been introduced and firmly mounted on the stage in order to ensure good thermal contact, the vacuum chamber must be sealed and pumped out with a rotary pump. Care must be taken at this stage not to prolong this initial pumping process as the vacuum chamber may gradually be filled with hydrocarbons originating from hot oil in the rotary pump, which may result in eventual contamination of the sample (Fig. 23).

This problem can be overcome by employing a turbo-molecular pumping system which produces an operational vacuum without the use of pumping fluids. This type of pump is often fitted to the Balzer apparatus. In the case of a conventional pumping system consisting of a diffusion pump backed by a rotary pump, a procedure must be adopted whereby the evacuation of the chamber is terminated when pressure in the roughing line of the system reaches approximately 10^{-1} Torr. At this pressure all the gas in the vacuum lines is still flowing towards the exhaust. At the same pressure (10^{-1} Torr) the diffusion pump can be safely connected to the chamber to evacuate the remainder of the gases. Of all residual gases in vacuum systems, water-vapour is the most persistent, and very difficult to pump out, mainly due to its strong affinity for metal surfaces. Removal of water-vapour can be considerably speeded up if a well-designed cooled baffle is superimposed between the chamber and the diffusion pump. When cooled with liquid nitrogen the baffle acts as an efficient cryo-pump which serves the double purpose of pumping water vapour from the chamber and stopping back streaming of oil originating from the diffusion pump. The efficiency with which the pumping system removes all the residual gases from the chamber varies according to the experimental conditions, size of pumps, pumping fluids and previous use of the apparatus. It is therefore, of considerable importance at some stage of evacuation to be able to assess the quality of vacuum and its adequacy for the safe fracturing of the sample. This requirement applies whatever apparatus is used.

Conventional Penning gauges do not on the whole give a precise indication of vacuum in the vicinity of 10^{-6} Torr. Therefore, it is preferable to use either an ionisation gauge, or a quadropole mass spectrometer gauge, as these can give a much better indication of the state of vacuum within the chamber in the region of 10^{-7} to 10^{-6} Torr. Also, a quadropole mass spectrometer gauge, in addition to measuring total pressure, can be tuned selectively to water-vapour, and in this way, measure its partial pressure. If however, these sophisticated indicators are not available, another very useful method for assessing vacuum conditions which have to be judged as adequate for fracturing of the sample may be employed. It involves the measurement of the time interval associated with a pressure rise between two arbitrary points on the Penning gauge scales when the chamber is isolated from the pump. By measuring this virtual leak rate, caused by outgassing of the equipment inside the chamber, a very good indication of the state of vacuum may be arrived at. After prolonged evacuation, a time interval as measured for the leak rate may be reached which will not be substantially improved even after prolonged evacuation. In practice therefore, at this stage of the procedure, fracturing of the sample may safely take place.

If it is not possible to reach such a previously established time period for the leak, rate the chamber should be checked for air or nitrogen leaks. Also, if carbon arcs are employed to produce the replicas, the carbon which is very porous and easily absorbs large amounts of gases and vapours, should be repeatedly heated up almost to the point of arcing in order to drive out the gases and vapours, which otherwise would be released at the very critical time when the sample is shadowed.

IV FRACTURE OF THE FROZEN SPECIMEN

In some commercially available apparatus like Balzer's machine, fracture of the frozen sample is effected by chipping small thin conchoidal flakes from it. This is achieved with precisely controlled movement of a cooled blade. On lowering of the blade, deeper regions of the frozen sample are revealed as the blade traverses the entire specimen. In addition to the useful fracture faces revealed by the removal of small chips, a percentage of the area is found to be covered with flat surfaces, where the blade scraped the specimen. The flat surfaces produced in this way are devoid of any useful information as friction between knife and ice destroys all features (Koehler, 1968). In addition the revealed portions from which the useful replica is eventually going to be produced, may be contaminated by detached ice fragments carried by the knife (Staehelin and Bertaud, 1971).

A different approach to fracturing has been adopted by Steere (1957) the originator of the technique, who cleaved the sample with a cold blade outside the vacuum chamber in a specially constructed box cooled with dry ice. Bullivant and Ames (1966) followed the ideas of Steere, substituting the cold box with a liquid nitrogen bath. In both cases the frozen and fractured sample is transferred into the vacuum chamber for further manipulation. The last method however, may lead to contamination of fracture faces by substances dissolved in liquid nitrogen (Moor, 1971).

The method of tissue fracturing devised in our laboratory (Stolinski, 1975a) avoids the use of blades and employs instead a specially devised cleaver mechanism (Fig. 12). The cleaver induces a crack at two opposite sides of the frozen specimen, fractures it, and finally sweeps the

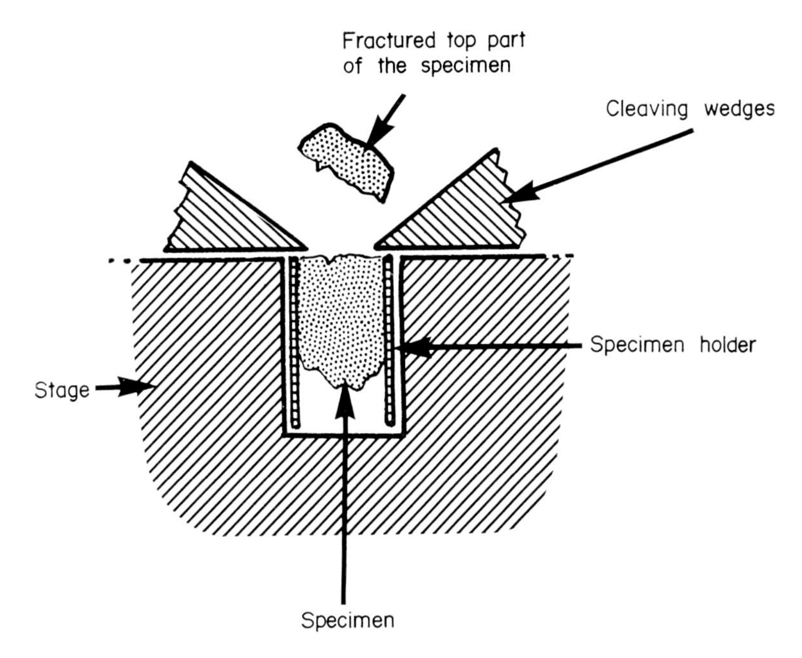

FIG. 12. Method of fracturing of tissue with a cleaver. Two converging wedges induce fracture in protruding specimen and then sweep the top portion of the fractured sample out of the way.

upper part of the specimen out of the way (Fig. 12). The main advantage of this arrangement is that the cleaver comes into contact only with the outer edges of the specimen, leaving its central part untouched. Samples may also be fractured by introducing them into hinged devices. However, on forcing such a device to open it is found that the surfaces tend to be much rougher when compared to those produced with a microtome blade or cleaver.

Finally it must be emphasised that the time period between fracturing and shadowing is the most crucial one as far as contamination of the sample is concerned. Therefore some form of shrouding of the specimen and the cleaving mechanism is very desirable. By maintaining such a shroud at low temperature a local clean vacuum is created and contaminants are trapped on the cold surfaces. Small openings in such a shroud can also effectively direct (collimate) the incoming shadowing material. In addition, the use of shutters which only open when the evaporation of shadowing material has already commenced can further reduce contamination, as during the warm-up period of the arc, when the shadowing material is not yet evaporated, other volatile substances and trapped vapours may be released.

A Matched Fractures

Basically, producing a matched pair of replicas from a specimen involves introducing them into a hinged device where the sample is conventionally frozen and eventually fractured by opening the hinge under vacuum (Steere, 1969). Various workers have constructed modified versions of such hinged devices. Hess and Bair (1972), for example, fitted such a device into the Balzer apparatus, and incorporated into it grease-smeared finder grids in order to guide the fracture plane between the grids.

In practice there are three main difficulties to overcome when

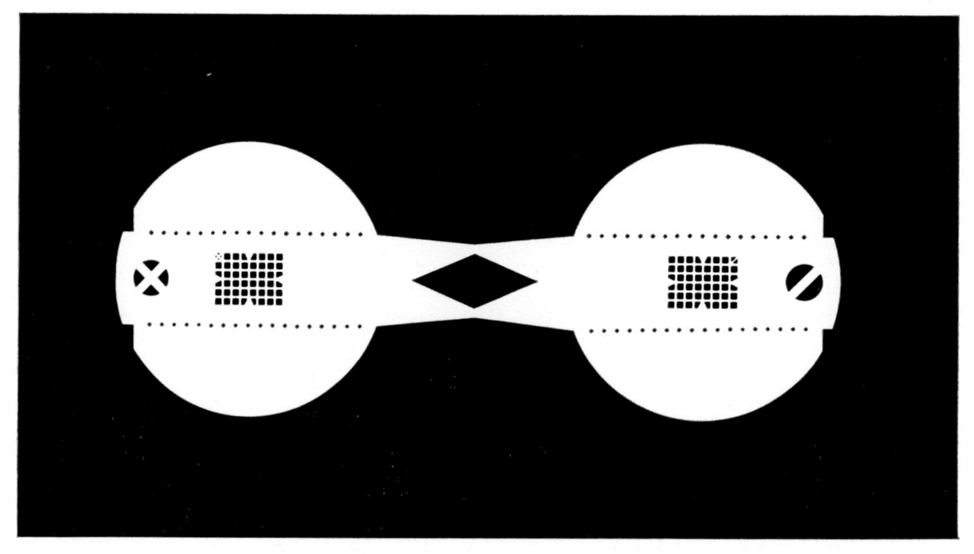

FIG. 13. Matched finder grids incorporating visual means for identification of co-ordinates on the fractured surface.

matched fracturing is undertaken. Firstly a good aid for identification of co-ordinates on the separate replicas has to be devised in the form of two matched finder grids. Unless such a device is available, the search for matched features can be very prolonged and frustrating. Secondly, the replicas have to be cleaned without being detached from the grids. This latter requirement is satisfied by using supporting grids made from a noble metal capable of withstanding the action of acids. Thirdly, a foolproof method for separation of the two finder grids must be devised. Preferable, this should be done without the use of grease which may be difficult to digest when cleaning the replicas. The grids, if not specially attached to the hinge device, may remain together after operation of the hinge mechanism, and if this happens the prime aim of the operation is not achieved. Many types of matched oyster grids are available on the market, but very few are suitable for the present purpose. Accordingly, a special pair of grids of a particular design for the identification of matched fracture faces has been designed in our laboratory (Fig. 13). This device, when folded, can easily be checked for correct matching of grid openings under the binocular microscope, and adjusted, if necessary, until they are lined up in both. Subsequently, the grids are clamped in the hinged mechanism and checked again for coincidence

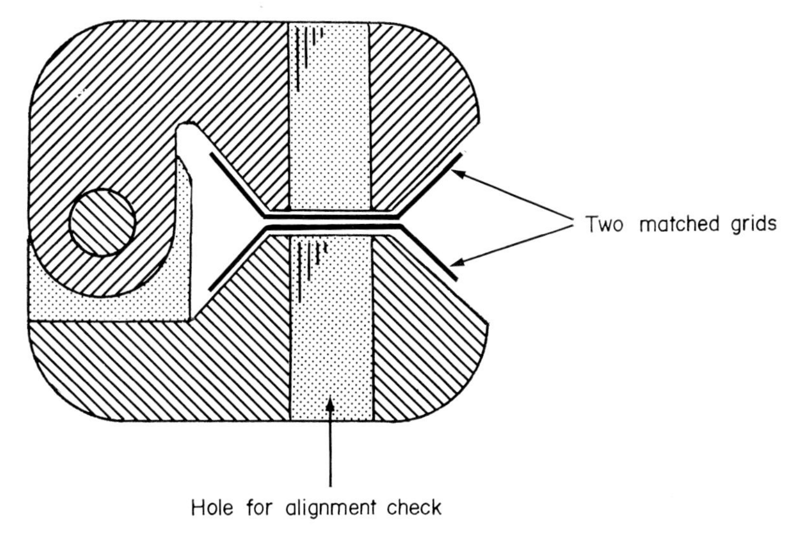

FIG. 14. Cross-section through a folded hinged mechanism for producing matched replicas. When operating the hinge mechanism the grids are retained on individual sides of the hinge, thus correctly guiding the fissure between the two matched grids.

(Fig. 14). In the case of suspensions, the sample can be introduced at this stage from both sides of the folded device (Fig. 14). In the case of tissue, small thin pieces can be introduced between the grids and then squashed so that the grids are almost in contact. As the amount of tissue is small the exact overlay of the two grids can still be checked by shining light through the tissue. The grids can be clamped at this stage in the hinged device. Prior to rapid freezing, the grids must be carefully wetted with the final solution of the cryoprotectant. This procedure, after rapid freezing, ensures strong attachment of the grids to the individual sides of the hinge device and their correct separation (Fig. 14).

Preferably the shadowing of the revealed surfaces should be such that after the hinged mechanism is operated, platinum-carbon is evaporated along the axis of the hinge. In this way identification of features on the replicas is achieved more easily when a pair of micrographs is presented side-by-side, as in such a case similar direction of shadowing is evident on both micrographs (Figs. 45 and 46).

V SUBLIMATION

The sublimation process requires a high degree of accuracy in the measurement of sample temperatures and a reliable reproduction of the same conditions from experiment to experiment. When it is the intention to reveal additional portions of the sample by sublimation, cryoprotectants other than glycerol must be employed (see p. 26), or a chemically fixed sample may be frozen in distilled water.

After conventional freezing, the sample is fractured at a temperature at which sublimation of ice occurs, i.e. around −100°C. In practice different types of apparatus will require slightly different temperatures to achieve the same result, as the geometry of the sample holder and the position of the thermometer element will affect the measurement of the actual temperature at the surface of the sublimating specimen. Provided therefore, that some preliminary experiments are made and the prevailing conditions are carefully noted, a temperature may be established which gives most satisfactory sublimation for a given time interval, preferably not exceeding 1 minute. It follows therefore that a precise method for the measurement of the temperature of the sample in the vicinity of −100°C is essential in addition to some means for controlling that temperature to at least ±1°C. When electric heaters are incorporated in the stage it is possible by gradual introduction of heat to the sample, to raise its temperature to the desired level and then hold it steady for the required time interval. The sublimation process can then be terminated by switching off the heater current which should result in lowering of the temperature to approximately −110°C at which stage of operation the replica may be formed.

Automatic devices are also available which can perform most of the described actions. For example by dialling the desired temperature the stage can be automatically heated or cooled in a proportional way.

VI REPLICATION

Shadowing of the revealed surfaces and subsequent evaporation of pure carbon can be carried out either with an electron gun or a carbon arc. In the case of an electron gun it is essential to eliminate energetic ions produced during evaportion which can inflict considerable damage on

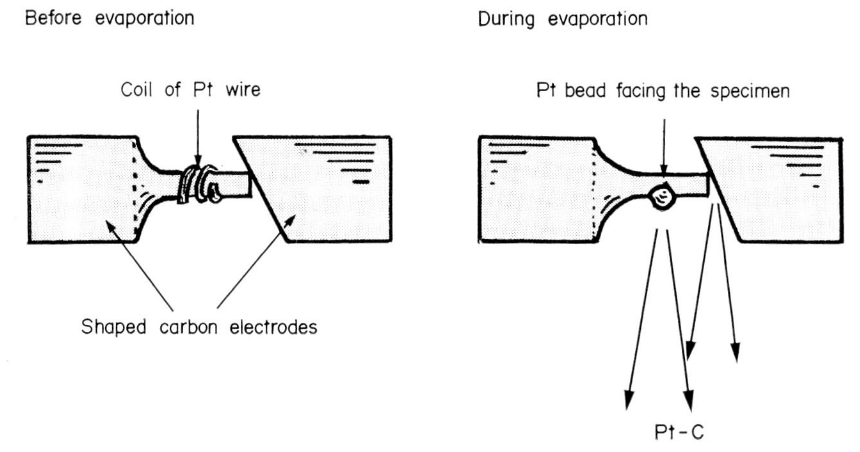

FIG. 15. Shape of evaporating electrodes. A length of platinum wire is wound on to the thinned portion of the carbon electrode and then melted under vacuum. When appropriately aligned the source may be used more than once.

the specimens (Moor, 1971). Electron guns are very reliable in operation and in the case of carbon evaporation, require a minimal amount of material, which can easily be de-gassed. The application of ion guns for evaporation is currently being investigated, as with this type of device evaporation can be performed at a low temperature with possibly less damage to the tissues. In the case of more conventional carbon arc evaporators every effort should be made to limit the length of the carbon rods in the electrode holders. In this way much more thorough de-gassing of carbon is possible, as then almost the entire bulk can be effectively heated. In freeze-fracture procedures complete de-gassing of the evaporation sources is of utmost importance. Badly de-gassed evaporation sources invariably contaminate freshly exposed surfaces with vapours and volatile substances, which at room temperature can be absorbed in large quantities by carbon. It must be realised that all these contaminants are released at temperatures which are not high enough for the evaporation proper to commence, resulting in contamination of freshly exposed surfaces.

For evaporation of platinum-carbon, a modified version of Moor's (1959) method has been adopted in our laboratory. A spiral of platinum wire (0·23 mm diameter, 20 mm long) is placed on an appropriately shaped electrode (Fig. 15). The platinum wire is then melted under vacuum and forms a small bead half-way along the length of the thinned carbon rod (1·5 mm diameter). If necessary, the entire rod may then be rotated, until the bead faces the specimen stage. In this way when evaporation is taking place, both carbon from the arc and platinum from the bead are simultaneously evaporated (Fig. 15). The two elements on arrival at the sample, form a uniform, fine grain high electron dense layer (Fig. 16b).

Good quality replicas can only be produced if the correct amount of shadowing material is evaporated on to the fractured specimen. Information is lost when too thin layers are evaporated while too thick layers reduce resolution due to piling up of material on the sample surface. A quartz-crystal film-thickness monitor provides one means of estimating that thickness. If this type of highly reliable but rather expensive device is not available some other type of thickness gauge has to be devised. This may take the form of a rectangular piece of thin PTFE sheet (15 × 30 mm), which on evaporation will darken, and act as a comparative thickness gauge. One half of the white sheet should be prevented from receiving the evaporant while still remaining visible to the observer. This can be achieved by shading part of the PTFE slab with a small piece of glass placed well in front of it. When evaporation takes place, intense light produced by the arc adequately illuminates both halves of the slab, so as to permit the observer to estimate the degree of darkening and hence the thickness of the film. The degree to

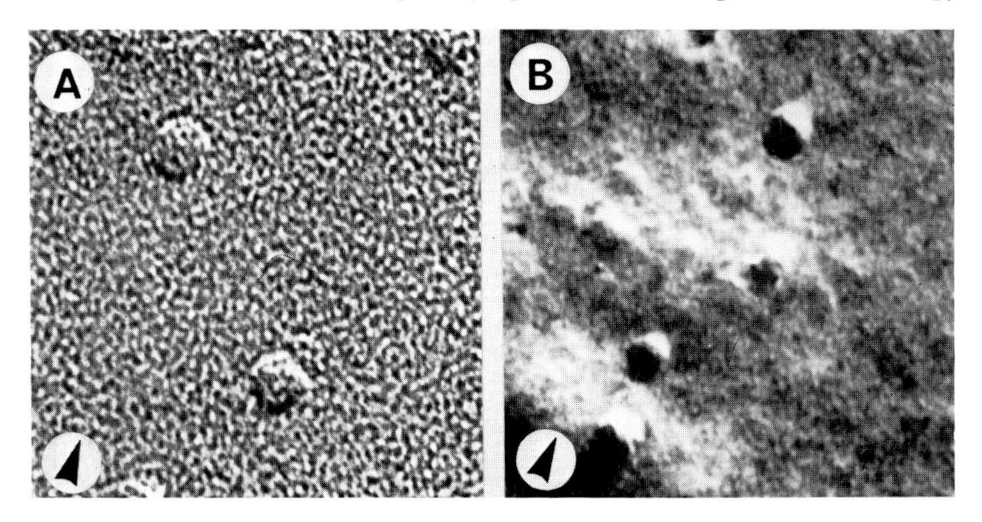

FIG. 16. Replicas showing membrane associated particles. In A, the particles are shadowed by pure platinum showing substantial grain size. In B, platinum carbon shadowing is seen to produce much more uniform grainless film. × 500,000.

which the white slab has darkened can then be further quantified by comparing it with standard density charts as supplied by photographic agencies.

An alternative way of producing uniform film thickness from experiment to experiment is by incorporating a timing mechanism with the evaporation source. Provided that the evaporation system is reliable and stable, some standard conditions may be set and the mechanism actuated for a pre-determined length of time. Nevertheless, however reliable the method of evaporation may be, some check of the thickness is most desirable for both platinum-carbon and carbon films. If possible the films should be evaporated in one continuous sequential operation with minimum interruption.

When pure platinum is evaporated from a source it tends to produce small grains on arrival at the shadowed surface. This is especially noticeable when platinum is evaporated from tungsten filaments or from a carbon arc which was not sufficiently tensioned. A small admixture of carbon to evaporated platinum results in a much more satisfactory grain-free film (Fig. 16). Also excessive amounts of evaporated pure carbon to bind together the platinum carbon must be avoided as resolution of features on the replica may be considerably impaired, when the carbon coat becomes noticeable on micrographs (Fig. 25).

Fractured specimens very often exhibit deep depressions and elevations and it is found that when the platinum carbon is evaporated at 45° to the plane of the specimen large areas of it remain in the shadow and are devoid of infomation. Because of this effect, it is found that an angle of approximately 60° is preferable, as then more of the specimen is revealed by the shadowing.

VII REMOVAL OF TISSUE FROM REPLICA

When the fractured sample with the formed replica is removed from the evaporator, it should not be thawed immediately. It should be placed temporarily in a small pool of liquid nitrogen, while the operator is attending to the evacuation of the evaporator chamber. The less this is open to the atmosphere the better vacuum conditions are going to be during subsequent runs.

The frozen sample in or on its container should next be gradually thawed by holding it above the liquid nitrogen bath, and then introduced into a small white porcelain dish. A porcelain container is advisable as small dark pieces of replicas are then easier to observe against the white background. In the dish, the sample should then be immersed in the original fluid with which it was finally permeated prior to rapid freezing. In this way osmotic forces leading to swelling or shrinking are minimised. With particularly difficult tissues which on thawing tend to shrink or swell excessively with resulting damage to replica it may be better to fix the sample in buffered gluteraldehyde which stabilises to a certain extent the tissue components.

In the case of cell suspensions no difficulties are normally encountered on thawing. With these the replica should, if possible, be floated on to the surface of distilled water to which nitric acid in the case of animal cells, and sulphuric acid in the case of plant cells, is gradually added to approximately 20% concentration. The strength of the acid acting on the tissue should then be increased by transferring the replica with either a platinum wire loop or pipette to another porcelain trough containing 100% of the relevant acid. Finally the replica should be transferred in all cases into fresh 100% nitric acid and heated in it to about 60°C for at least 25 minutes. Care must be taken at this stage to ensure that the acid does not evaporate completely from the dish, thus leaving the replica dry. The most convenient heater for the acid bath is the flat top of the lamp shade as used for illuminating objects observed under a binocular microscope.

For soft organised tissues, like liver, the following modified procedure is recommended:

1. Gradually thaw the sample.
2. Introduce holder with tissue into original cryoprotectant.
3. Slowly increase the percentage of nitric acid up to 100% to digest tissue for 1 hour.
4. Heat the tissue in 100% strong acid to 60°C for about 25 minutes.
5. Choose one small fragment of replica and wash it twice in distilled water.
6. Pick up the replica on a grid and examine under the electron microscope. If not clean, heat the original replica again to 60°C for 20 minutes and examine a second small fragment. If clean, the whole replica may then be picked up.

Basically four types of reagents are used according to their efficacy in digesting tissues:

1. Nitric acid up to 100%.
2. Sulphuric acid up to 100%.
3. Caustic soda up to 40%.
4. Sodium hypochloride solution with 10–14% available free chlorine.

All the above solutions have the advantage of being transparent, thus making the replica fragments visible against the white porcelain background. In practice the digesting fluids are chosen according to type of tissue and its constituents. When tougher tissues are processed very gradual digestion is recommended. For example, leaving the tissue over the week-end or longer in weak acid very often produces best results.

When picking up the cleaned replica, it is advisable to choose a grid with a mesh which is capable of adequately supporting it. For example, extensive replicas can be examined on grids with 50 × 50 μm openings. In removing the replica from the bath, the grid should approach it from below, and this operation should be carried out under a binocular microscope. After drying, the replica is ready for examination in the electron microscope. If the fragments of replica are so small that the grid mesh is not capable of supporting them, a grid coated with a carbon film may be used. However, in most cases, uncoated grids are suitable.

VIII ARTIFACTS

As with any technique involving a series of processing steps or manoeuvres, freeze-fracture is prone to various artifacts, which will appear as defects on the final electron micrograph. These may be considered as being caused by:

1. Autolytic or necrotic changes occurring within the tissues while they are being permeated with cryoprotectant (Fig. 17). This type of artifact may be minimised by chemical pre-fixation.

2. Temperature change, when permeation of tissue is carried out at temperatures significantly lower than physiological. The resulting artifact involves redistribution or aggregation of particles associated with membranes (Fig. 18), leading to mosaic patterning and the occurrence of smooth areas. This type of artefact may also result from, or be indicative of a degree of necrosis (McIntyre *et al.*, 1974). In this connection it must be emphasised that absence of particles from an area of membrane is not necessarily artifactual; it is normal with myelin (Fig. 101) and with stratum corneum plasma membrane (Fig. 40). However, when this feature is observed on other material, it is advisable to check for collateral evidence of necrosis or temperature effect.

3. Formation of ice crystals within the tissues when an inadequate amount of cryoprotectant has been employed. The tissue may be completely distorted (Fig. 19), or, the damage may be recognisable only at high magnification, depending upon the size of the ice crystals.

4. The actual process of fracturing. This can lead to the production of furrows and steps which do not represent true morphological features (Fig. 20), or to plastic deformation of structures which may be pulled out of the ice matrix before they fracture (Fig. 72).

5. Deposition of extraneous material on the fracture face of the specimen before shadowing and replication has been performed. This can include:

(*a*) Deposition of intermediate cooling agents such as Isopentane or

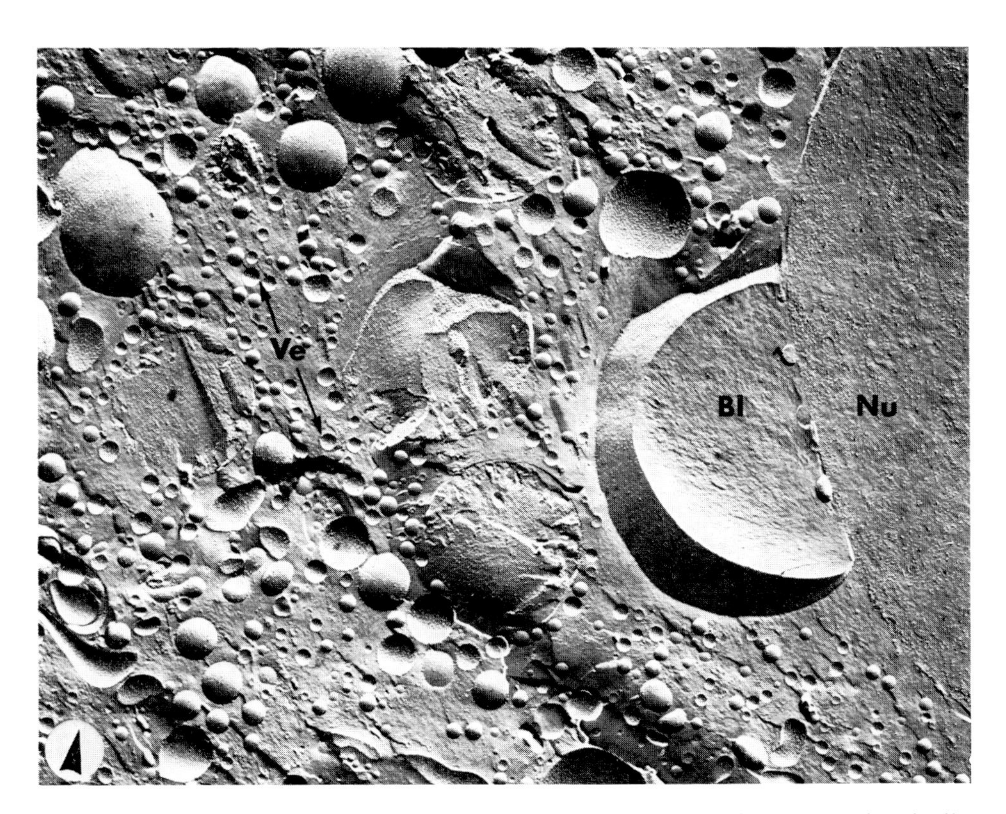

FIG. 17. Replica of parenchymatous liver cell of mouse which was not chemically fixed prior to permeation with cryoprotectant. The bleb (Bl) extending from the nucleus (Nu) is thought to be a necrotic feature, and the uniformly sized small vesicles (Ve) have been interpreted as resulting from break-up of rough endoplasmic reticulum. It should be emphasised that such necrotic manifestations are not invariably seen in all tissues not pre-fixed chemically. × 17,000. In this and in all subsequent micrographs, direction of shadowing is indicated by arrow at bottom left. All illustrations in this book are of material which was not exposed to chemical fixation.

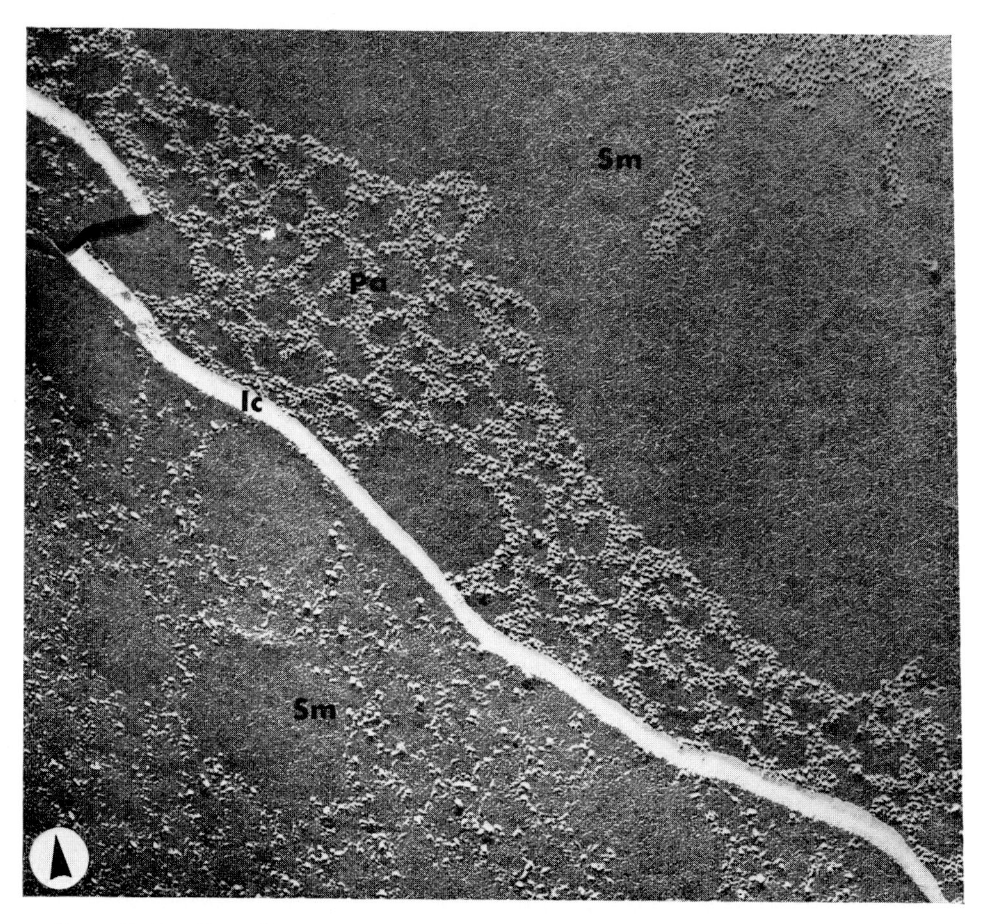

FIG. 18. Replica of membranes of cells of mouse retina which was kept at 0°C during permeation with cryoprotectant. Note aggregation and mosaic arrangement of particles (Pa) leading to extensive smooth areas (Sm) on the membranes. Ic, intercellular material. × 70,500.

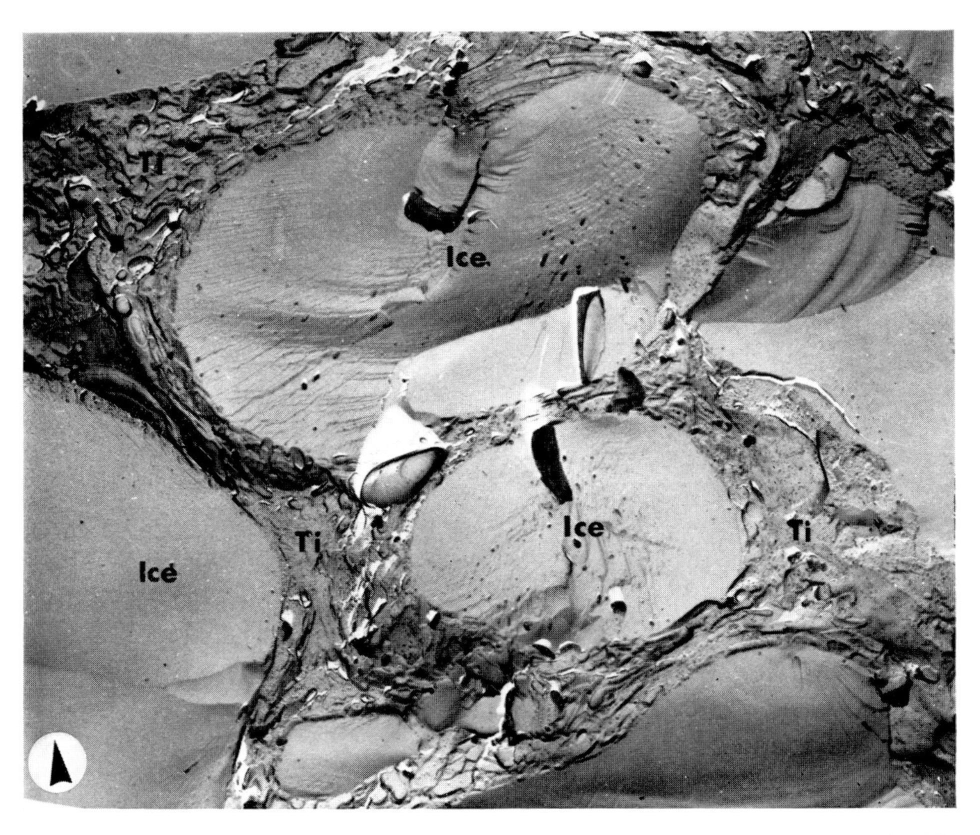

FIG. 19. Ice crystals distorting kidney tissue (Ti) inadequately cryoprotected with glycerol. × 18,500.

FIG. 20. Replica of cell cytoplasm showing steps (St) produced by stresses occurring during fracturing. These do not represent true morphological features, Ti, undigested tissue contaminating the replica. $\times$ 46,500.

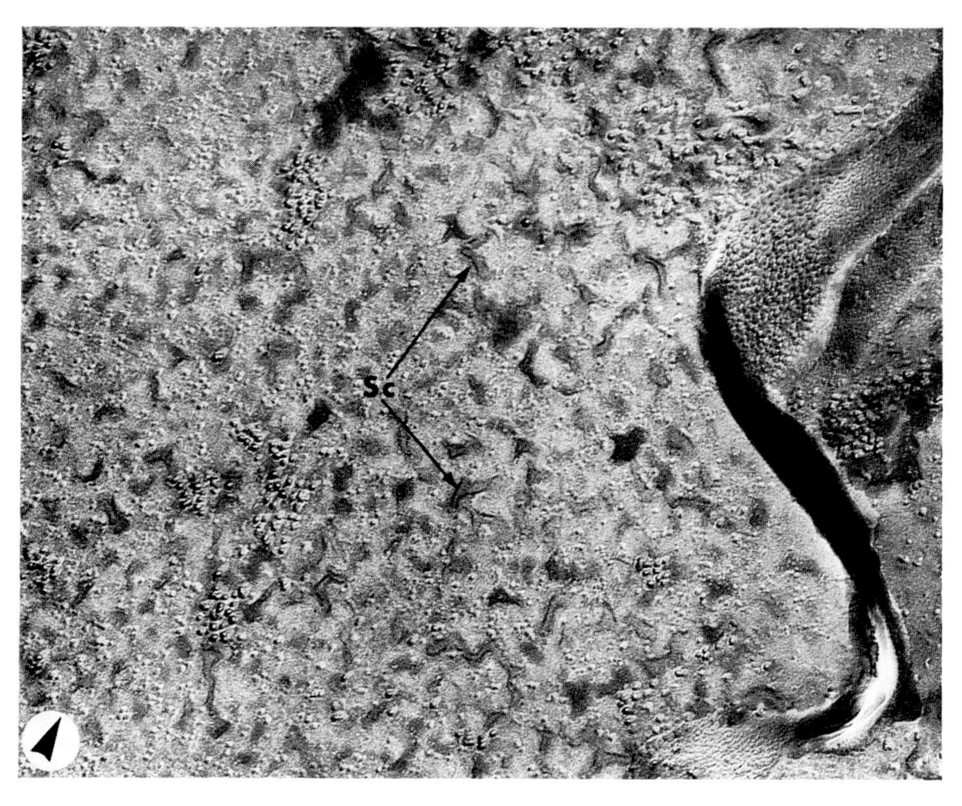

FIG. 21. Replica of epidermal keratinocyte contaminated with flakes or scales (Sc) of Isopentane. × 80,000.

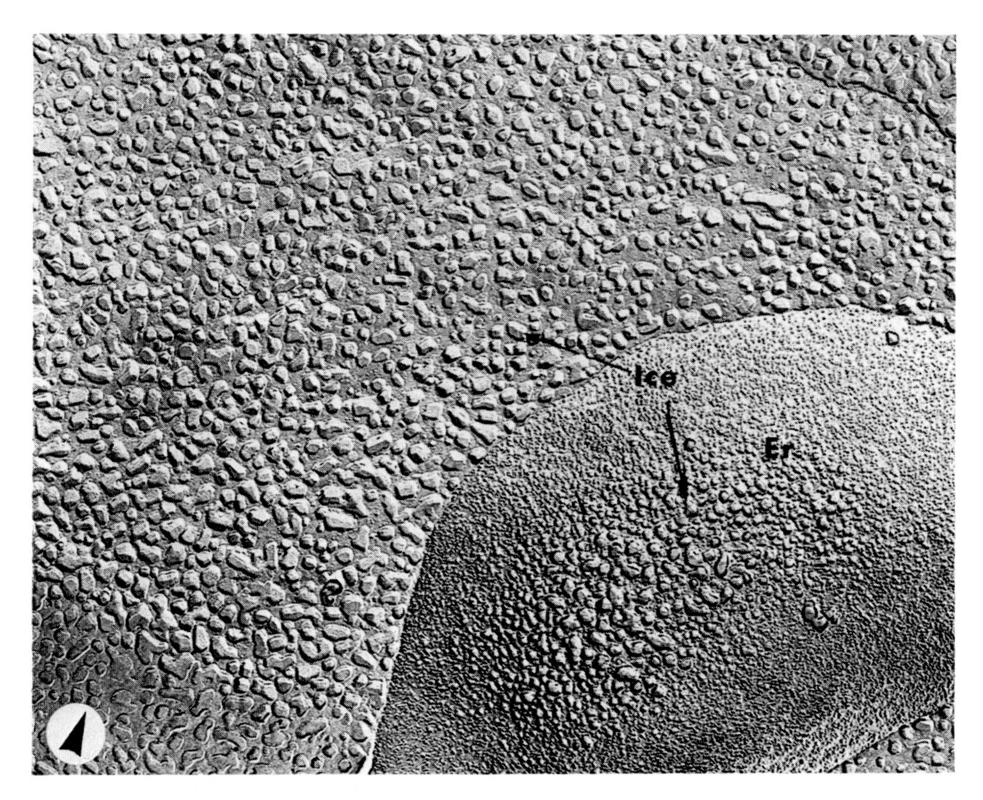

FIG. 22. Replica of erythrocyte (Er) showing ice crystals formed in the interval during fracturing and shadowing due to a vacuum air-leak. Ice is present both on the fracture face of the erythrocyte membrane, and on the fractured medium. × 35,700.

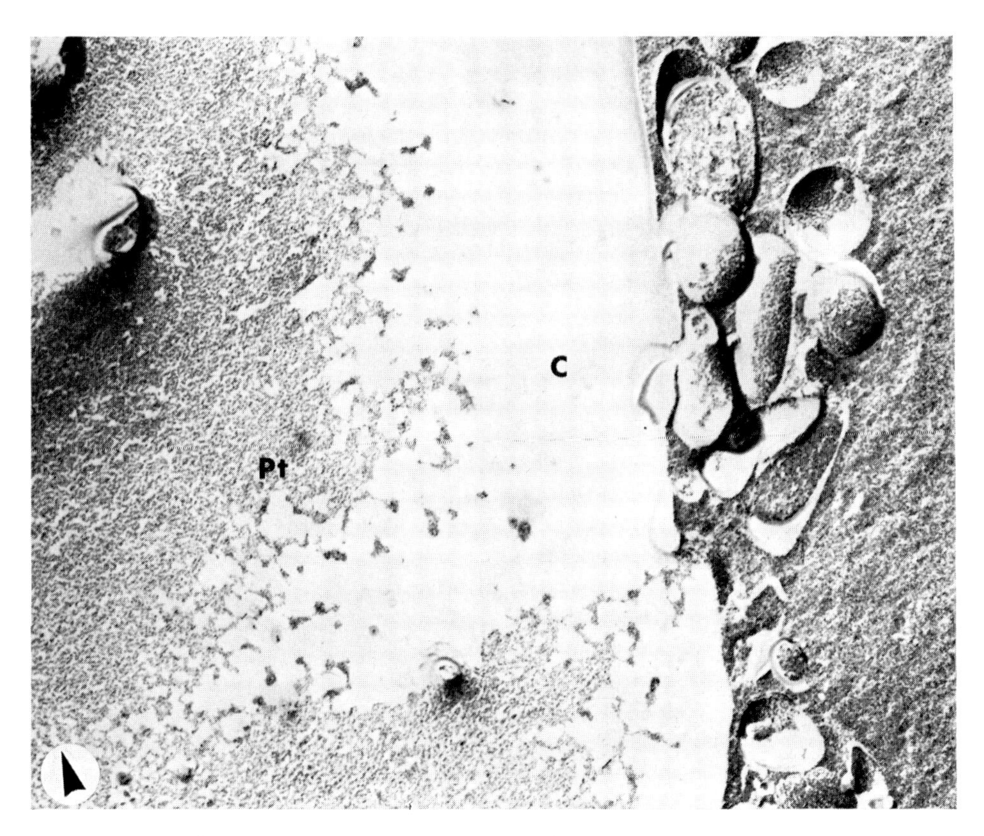

FIG. 23. Replica showing uneven deposition of platinum-carbon (Pt) due to deposition of diffusion pump oil on the fractured specimen. C, area of replica consisting of pure carbon only. × 52,500.

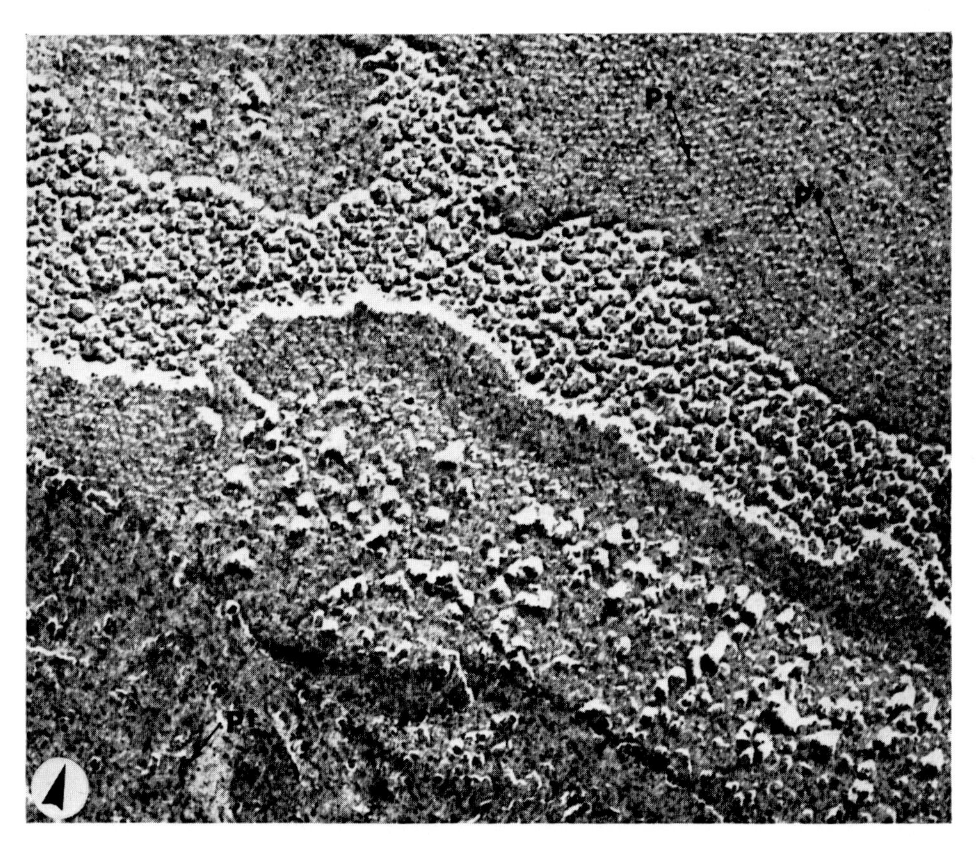

FIG. 24. Replica showing 'black-spotting' contamination due to crystallisation of platinum into relatively large grains (Pt) following intense electron bombardment of the replica in the microscope. × 187,600.

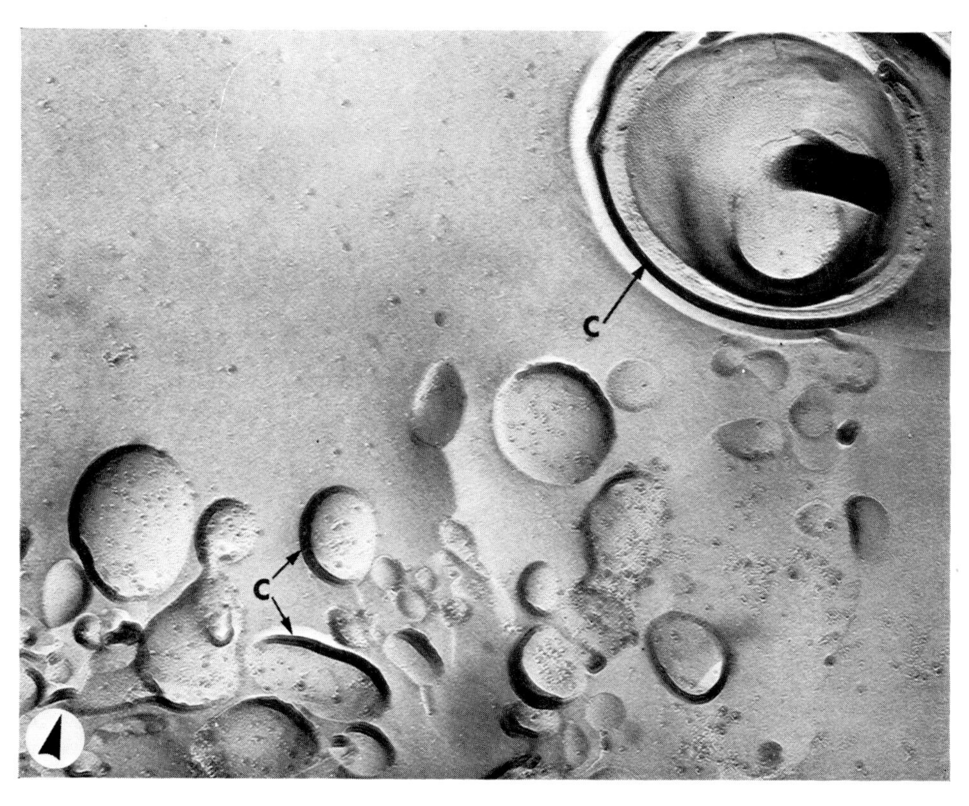

FIG. 25. Replica of cell organelles inadequately coated with platinum carbon, and exhibiting regions where carbon (C) appears as a dense broad line when viewed vertically. × 56,000.

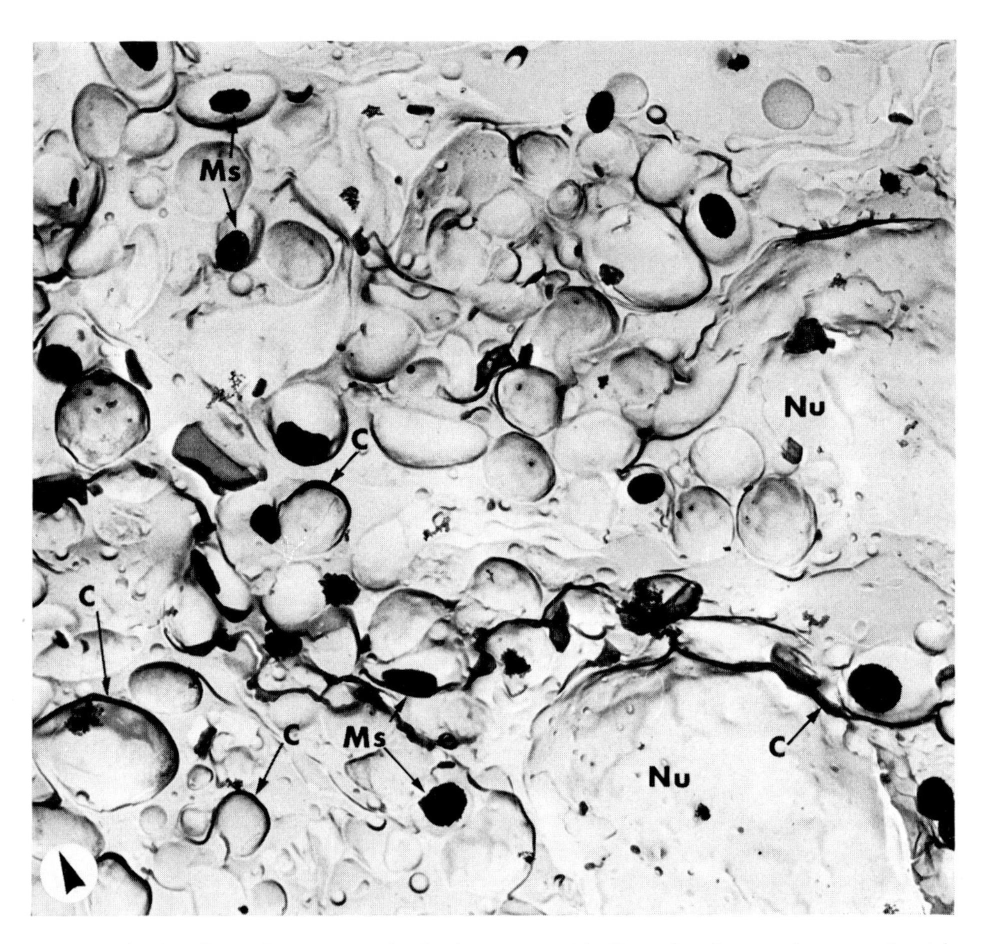

FIG. 26. Replica of mouse retinal pigment epithelium inadequately coated with platinum carbon, and exhibiting 'carbon line' effect (C). Ms, undigested melanosomes attached to replica; Nu, nucleus. × 7000.

Freon, which may freeze on the specimen in the form of a crust which, after fracturing may sublimate, and then re-freeze on the exposed fracture surface (Fig. 21). This can be avoided by removing excess of frozen coolant from the specimen before introducing it into the vacuum chamber (see p. 37), and by temporarily raising the temperature of the specimen to −140°C for a few minutes prior to fracturing, thus ensuring removal of intermediate coolant by sublimation.

(*b*) Deposition of ice-crystals in the presence of an air-leak into the vacuum chamber (Fig. 22).

(*c*) Deposition of diffusion pump oil (Fig. 23).

6. Excessive accumulation of shadowing material on small objects. This can lead to distortion of original shapes, e.g. the filling up, or obliteration of small holes.

7. Crystallisation of platinum to relatively large grains due to heavy electron bombardment in the microscope leading to 'black-spotting' of the final print (Fig. 24).

8. Excessive deposition on the platinum-carbon, of the pure carbon which binds the replica. Despite the fact that carbon is essentially transparent to electrons, it may be of sufficient thickness, especially when observed vertically or tangentially, so as to register as a solid coat covering particular features (Fig. 25). This effect is enhanced if, at the same time, too little platinum was evaporated, leading to poor general contrast.

9. The presence of incompletely digested tissue on the under surface of the replica (Fig. 26).

Most of these artifacts can be avoided by taking appropriate preventive action.

REFERENCES

Alink, G. M., Verheal, C. C., Offerijns, F. G. J. (1974). *In* 'Abstracts for Society of Cryobiology'. 11th Annual Meeting. London, Abstract 31.

Bachmann, L. and Schmitt, W. W. (1971). *Proc. Nat. Acad. Sci.* U.S.A. **68**, 2149–2152.

Bachmann, L. and Schmitt-Fumian, W. W. (1973). *In* 'Freeze-etching, techniques and applications'. (E. L. Benedetti and P. Favard, eds) 73–79. Société Française de Microscopie Électronique Paris.

Branton, D. and Moor, H. (1964). *J. Ultrastruct. Res.* **11**, 401–411.

Bullivant, S. and Ames, A. (1966). *J. Cell Biol.* **29**, 435–447.

Dempsey, G. P., Bullivant, S. and Watkins, W. B. (1973). *Science* **179**, 190–192.
Hess, W. M. and Bair, R. L. (1972). *Stain Techn.* **47**, 249–255.
James, R. and Branton, D. (1971). *Biochem. Biophys. Acta* **233**, 504–512.
Koehler, J. K. (1968). *Adv. Biol. Med. Physics* **12**, 1–84.
MacKenzie, A. P. (1969). *Biodynamica* **10**, 341–351.
McIntyre, J. A., Gilula, N. B. and Karnovsky M. J. (1974). *J. Cell Biol.* **60**, 192–203.
Moor, H. (1959). *J. Ultrastruct. Res.* **2**, 393–422.
Moor, H. (1971). *Phil. Trans. Roy. Soc. Lond.* **B261**, 121–131.
Nanninga, N. (1973). *In* 'Freeze-etching, techniques and applications'. (E. L. Benedetti and P. Favard, eds) 151–179. Société Française de Microscopie Électronique, Paris.
Speth, V. and Wunderlich, F. (1973). *Biochim. Biophys. Acta* **291**, 621–628.
Staehelin, L. A. and Bertaud, W. (1971). *J. Ultrastruct. Res.* **37**, 146–168.
Steere, R. L. (1957). *J. Biophys. Biochem. Cytol.* **3**, 45–60.
Steere, R. L. (1969). *In* '27th Ann. Proc. E.M.S.A.' (C. J. Arcenaux, ed.) 202–203.
Stolinski, C. (1973). *Proc. Roy. Mic. Soc.* **8**, 191–192.
Stolinksi, C. (1975). *J. Micros.* **104** (in press).
Umrath, W. (1974). *J. of Micr.* **101**, 103–105.
Wunderlich, F., Speth, V., Batz, W. and Kleinig, H. (1973). *Biochim. Biophys. Acta* **291**, 621–628.
Wunderlich, F., Batz, W., Speth. V. and Wallach, D. H. (1974). *J. Cell. Biol.* **61**, 633–640.

4

Interpretation of Replica Images in Micrographs

I THEORETICAL INTERPRETATION OF IMAGES

When the appearance of a surface is examined by the naked eye, or with a light microscope, it is seen by reflected light and in the case of the scanning electron microscope, by scattered electrons. The transmission electron microscopist, in order to observe such a solid surface, has had to devise means of transferring its details on to a thin electron translucent faithful facsimile, or replica. The thin film which makes up a replica presents to the electron beam local variations in thickness of evaporated material which, through electron scattering, produce a faithful representation of the surface detail of the original. The technique was first successfully demonstrated by Shaefer and Harker (1942) who photographed a replica of an irregular surface made from a plastic film on to which shadowing metal was evaporated from an angle. This latter procedure puts into relief the topography of the replica, because surfaces facing normally to the source of evaporation will accumulate more material than others subtending a lower angle to it, so that when the replica is photographed, areas of different densities will be seen in the print. If the direction of metal shadowing is known, it becomes possible to identify from the varying densities observed, if, for example, a step seen is up or down. Similarly, partly uncovered spherical objects can be identified either as elevations or depressions. With current freeze-fracture practice, as explained in Chapter 3, the shadowed replica is produced by simultaneous evaporation of platinum and carbon, followed by further carbon evaporation to bind the original replica into a unified thin film.

A simple analogy can be made between the replica of a specimen and an undulating surface covered with a transparent detachable film. Figure 27 represents such a model which was sprayed from an angle

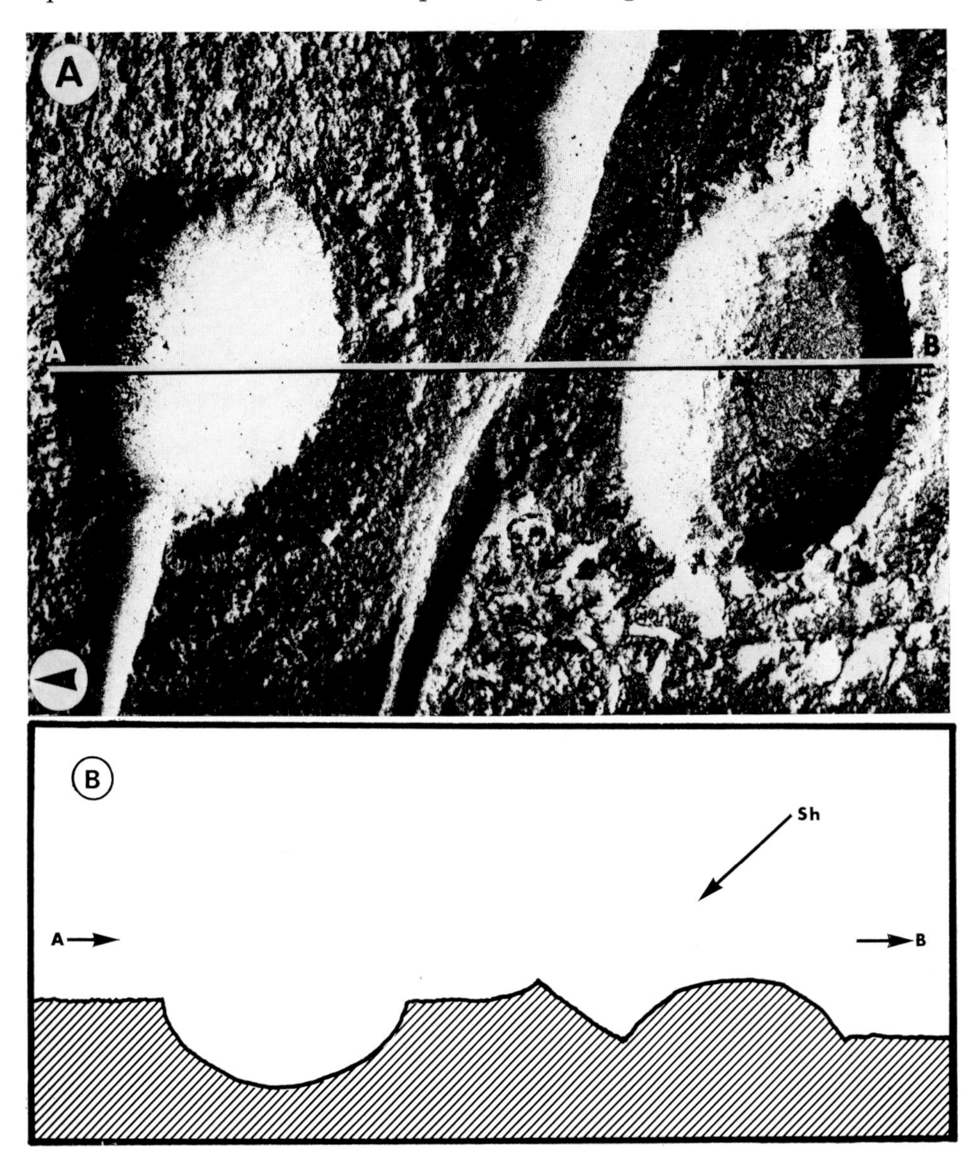

FIG. 27. A, Photograph of a replica model sprayed from an angle with black paint. B, sectional profile along the line A–B in A above. Sh, direction of shadowing.

with black paint. Provided that the direction from which the sprayer was aimed is established, the observed shapes may be perceived in three dimensions from the amount of black paint accumulated to a varying extent on the model surface. Those parts of individual features which face the source of paint (or in the case of a replica, the source of shadowing material) accumulate maximum amounts of shadowing material while other areas receive less, or none at all. When the thin film with the accumulation of paint is subsequently lifted from the model and examined by transmitted light it is observed under conditions analogous to those obtaining during examination of a shadowed replica in the electron microscope. The varying absorption of light over the surface of the film as caused by the uneven accumulation of black paint, can be interpreted by an experienced observer as true three-dimensional topography. The difficulty to the observer who is unfamiliar with such images (especially when presented on a photograph) is the correct perception of three-dimensional objects from a two-dimensional flat image. It is necessary for example to decide which part of the observed image represents an elevation and which a depression.

In order to explain how an analysis of such images can be achieved, diagrams as presented in Figs. 28 and 29 can be produced. In preparing them it must be assumed that, in the case of replicas, only small amounts of shadowing material are considered, which result in very thin layers. In that case when the replica is illuminated with electrons, the density presented to the observer is directly proportional to the thickness of evaporated material. Therefore, for the purpose of analysis it is necessary to arrive at some estimation of that thickness as presented to the source of illumination. It is also assumed that the evaporated material on arrival sticks to the examined surface.

For example, when a diagram is drawn of a convex hemi-cylinder lying on a flat surface, with its axis perpendicular to the source of the evaporating material, a density profile as on Fig. 28 is observed. However, when the same type of graphical analysis is applied to a hemi-cylindrical groove of the same dimensions, under similar conditions a different density profile, as shown and explained on the same Fig. 28, is produced. In practice it is observed that these graphical results are confirmed when for example a matched replica of a spherical object is examined under the microscope (Fig. 30). In the case of a depression, a shadow terminated half-way along the width of the object

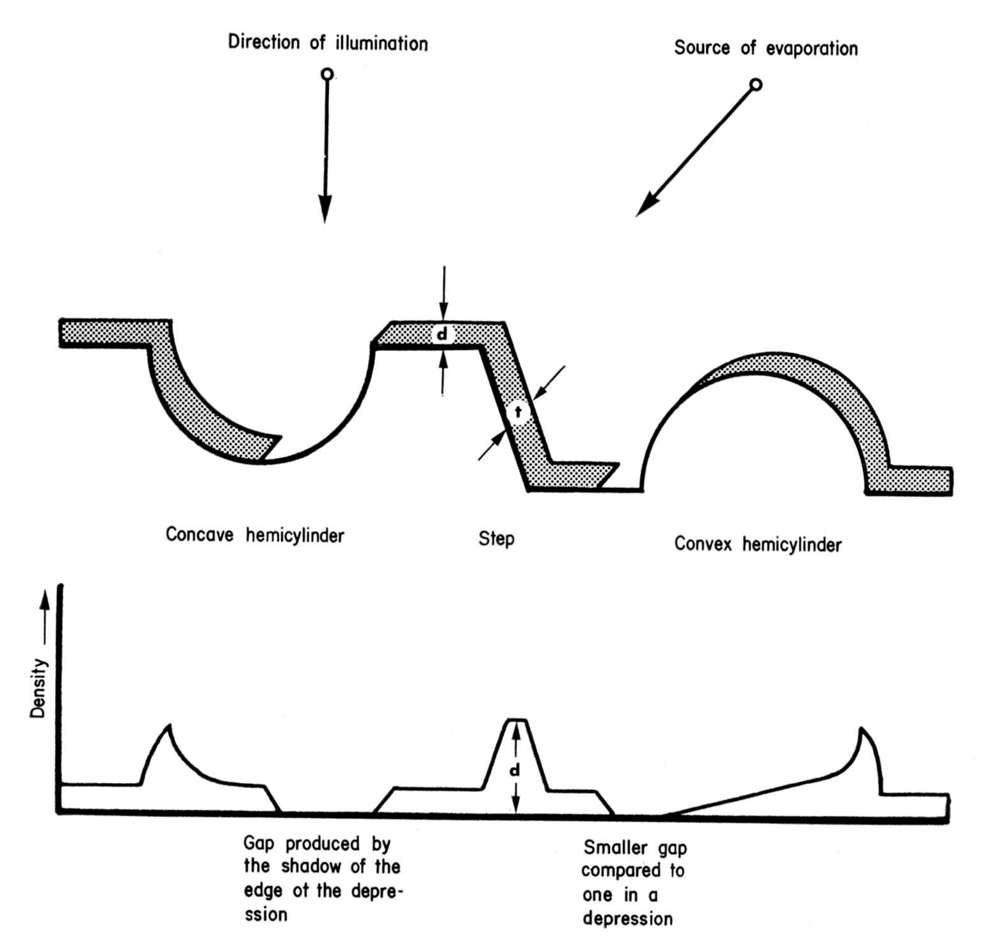

FIG. 28. Drawing showing theoretical density profiles of typical objects commonly observed on replicas. Note differences in profiles as produced by similar but inverted hemi-spherical shapes. t, thickness of the evaporated shadowing material measured in the direction towards the source of evaporation; d, thickness of the evaporated shadowing material seen in the direction towards the illumination source.

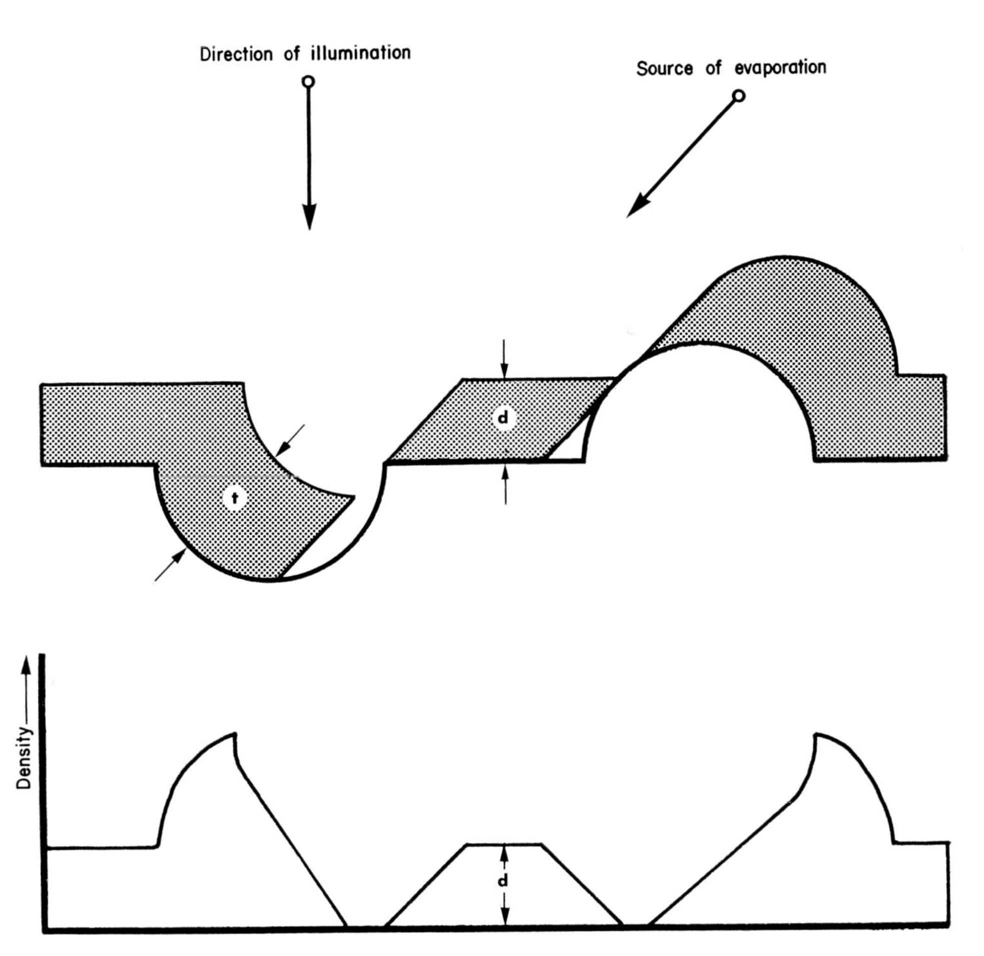

FIG. 29. Drawing of theoretical density profiles of small objects whose dimensions are approximately of the same order as the thickness of the evaporated film. Note that density profiles are mirror images of each other in such cases. t, thickness of the evaporated shadowing material measured in the direction, towards the source of evaporation; d, thickness of the evaporated shadowing material as seen in the direction towards the illumination source.

is observed, while in the case of an elevation the shadow extends beyond the object, and the material accumulated on its top extends approximately three-quarter way along its length. Therefore in the case of a hemi-spherical depression a shadow in the shape of a hemi-circle is produced which faces with the curved part, in the direction of evaporation, while in the case of a hemi-spherical elevation a crescentic shaped shadow is produced which faces the evaporation source with the inner part of the crescent, or its two sharp corners.

When a comparable analysis is undertaken for similar but much smaller objects which are of approximately the same magnitude as the thickness of the evaporated film, the resulting shadows are not so clearly diversified as in the previous case (Fig. 29). The density profiles produced are similar in shape and the only feature which distinguishes them is that they are mirror images of each other. That is, depressions show an accumulation of shadowing material on the side opposite to that of the accumulation on comparable elevations. When analysing such features along the direction of shadowing therefore, it should be remembered that in the case of a small elevation an accumulation of

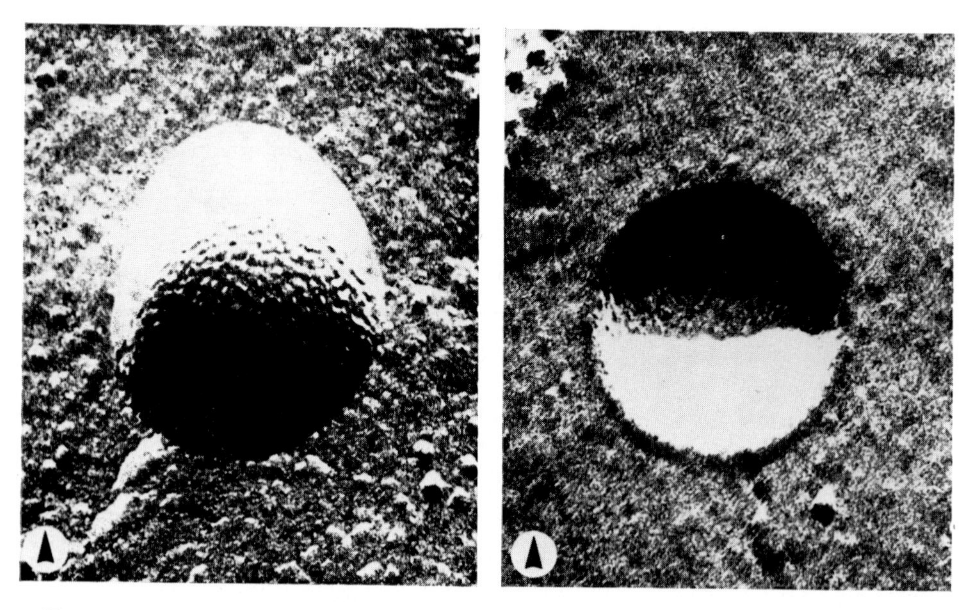

FIG. 30. Matched replica of a spherical object illustrating distribution of shadowing material on a hemi-spherical elevation (a) and depression (b). Compare with theoretical profiles shown on Fig. 28. × 216,000.

shadowing material is observed which is followed by a shadow; in the case of a small depression, a short shadow is followed by accumulation of material on the inner edge of that depression. For a more exact mathematical analysis of such images a method provided by Calbick (1951) can be also employed. However, the constructed diagrams can also be employed usefully to draw a number of guidelines regarding distribution of evaporated material on diverse shapes and their correct identification.

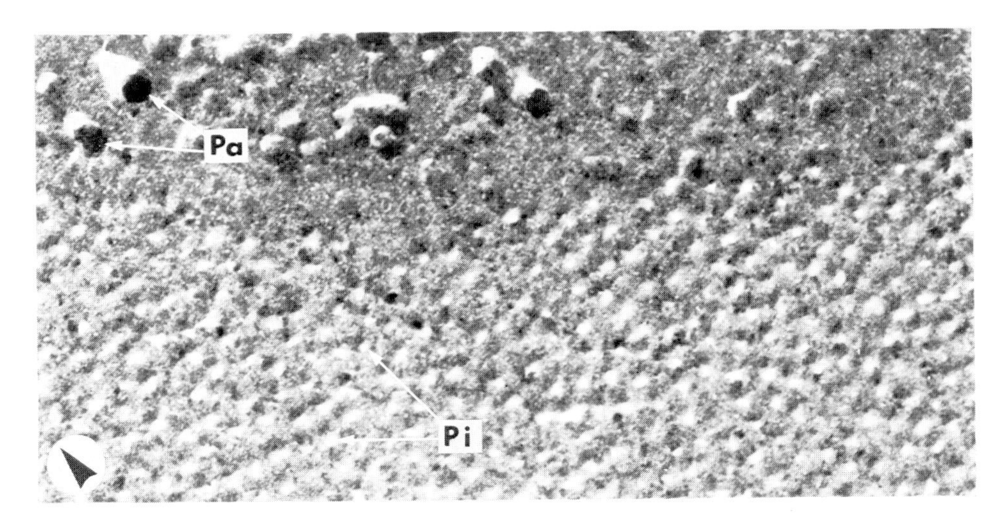

Fig. 31. Replica exhibiting small pits associated with the gap junction. Accumulation of shadowing material at small depressions (Pi) and particles (Pa) can be compared with the theoretical profiles shown on Fig. 29. × 501,000.

In practice many workers experience difficulties in demonstrating convincingly small depressions. In this context it must be appreciated that a very small amount of contamination may obliterate such features. However, under favourable conditions, (Fig. 31), very small depressions which are present on fracture faces associated with membranes can be readily demonstrated.

A Resolution of the Method

The above considerations lead to the theoretical estimation of resolution of the method. If, for graphical analysis, steps of equal height are chosen (Fig. 32), with their axis perpendicular to the source of

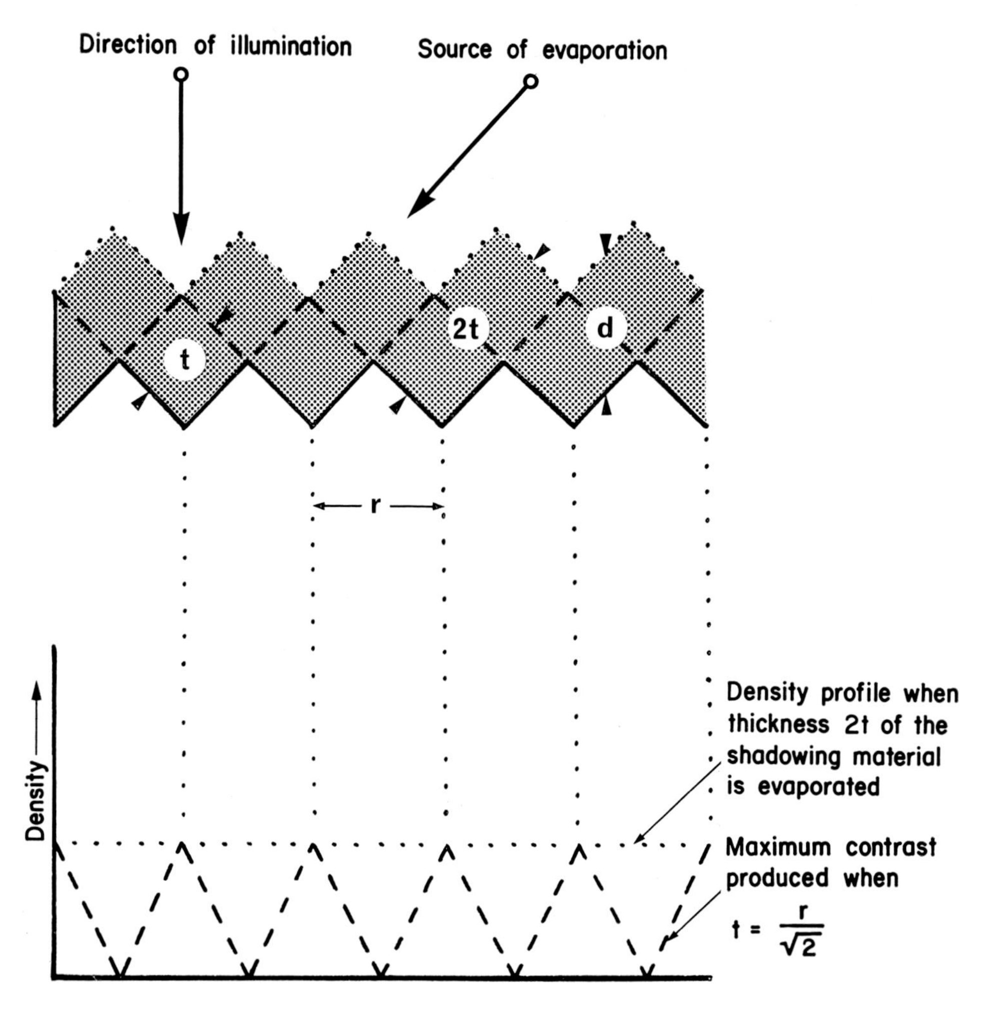

FIG. 32. Theoretical density profiles produced by steps of equal height. t, thickness of the evaporated shadowing material; d, thickness of same material but observed in the direction of the illumination source; r, possible resolution.

evaporation and lying normally to the illumination source, it is seen that optimum resolution is achieved when $r = t\sqrt{2}$. The steps are then seen under ideal conditions at maximum contrast, and are shifted by a distance equal to $r/2$ towards the source of evaporation. Also, it can be seen that when 't' increases to $2t$, the pattern vanishes as also demonstrated on Fig. 32. Therefore if for example, a 2·5 nm thick platinum coat is evaporated on to a stepped test object, it should be possible to resolve $2{\cdot}5 \times \sqrt{2} = 3{\cdot}5$ nm spaced steps. Figure 33 illustrates a

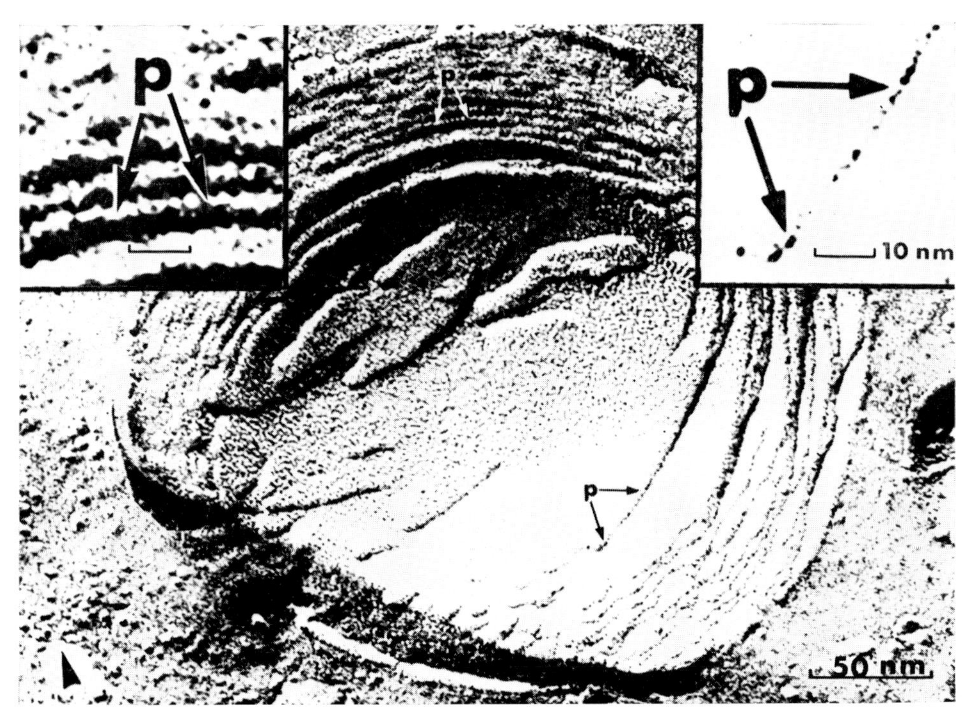

FIG. 33. Replica of fractured fat droplets exhibiting steps with particles (Pa) with 2·5 nm separation. × 330,000. Insets, × 960,000.

fractured globule of fat with concentric layers of lipids outlined by steps. When an approximately 2 nm thick layer of platinum carbon is evaporated on such a surface, a resolution of approximately 2 to 3 nm can be expected.

It follows therefore that the ability to perceive small objects on the micrograph does not depend solely on the grain size of the shadowing material but also on the shape of the surface under examination. However, the shadowing material on arrival at the examined surface may undergo scattering thus accumulating in areas which were originally shaded from the evaporation source. The same evaporated material may also tend to reform on the examined surfaces, resulting in production of large grains, instead of a reasonably uniform film. All of these effects may combine and effectively reduce resolution depending on the severity of each individual effect. In practice it is found that under optimal conditions resolution of the method is in the region of 2 to 3 nm. Provided therefore, that clean vacuum conditions are produced which minimise contamination, and shadowing material does not show a gross tendency to scatter and nucleate on the specimen surface, it should be possible routinely to achieve such resolution.

II PRINCIPLES UNDERLYING ANALYSIS OF MICROGRAPHS

It has become generally accepted practice to print micrographs of replicas directly from the original negatives, so that accumulations of evaporated metal appear dark on the print, while shadows appear as lighter areas.

A Determining Direction of Shadowing

In order to be able to interpret in three dimensions the topography of a replica as recorded on an electron micrograph, it is necessary, first of all, to determine from which direction the platinum-carbon or other shadowing material was evaporated on to the fractured specimen. Usually this is possible, especially with higher power micrographs, by locating a small isolated particle with its associated shadow. The shapes and orientations of shadows, preferably confirmed at different areas on the print, can then give an accurate indication of the direction of shadowing (Fig. 34). Examination of shadows related to spherical objects can also be of assistance in indicating direction. As explained previously (p. 68), shadows cast by convex objects tend to be crescentic in shape with the convexity of the crescent directed away from the shadowing source; shadows associated with concavities tend to be hemi-spherical with a sharp linear border. By studying the appearance of such shadows, one can decide which objects are convex or concave, and thereby

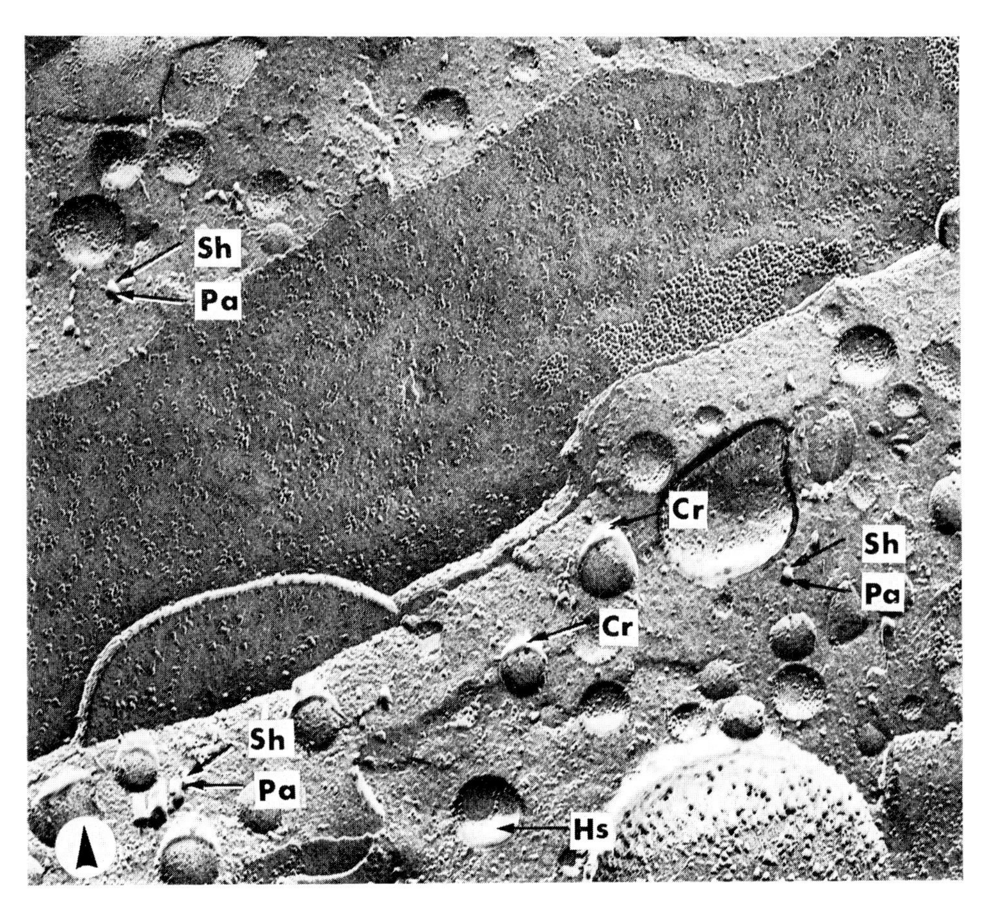

FIG. 34. Determining the direction of shadowing. In this micrograph of a replica of mouse liver, shadows (Sh) of particles (Pa) are seen to point away from the bottom of the print, in the direction indicated by the large arrow at lower left. Note shape of shadows (Cr, Hs) associated with spherical organelles in the cytoplasm. As explained in the text, examination of such shadows can also indicate the direction from which shadowing material was evaporated on to the fractured specimen. × 56,000.

deduce the direction of shadowing. With low-power, or survey micrographs, it may be more difficult to be certain of the direction of shadowing, though the occasional presence of a speck of contaminating material, or tissue fragment dislodged from one area and deposited on another during fracture may indicate it, by virtue of the prominent shadows cast in these circumstances. It is the convention to present micrographs as if the direction of shadowing (indicated by large arrow) were from the bottom of the print, and this practice is adhered to, as far as possible, in this book.

B Fracture Faces and other Features

A typical electron micrograph of a replica of freeze-fractured unsublimated biological material reveals two basic types of fracture face (Fig. 35). One is a *rough* randomly produced fracture of the composite material situated inside or outside the cell. The composite materials in this case will be respectively, the cytoplasmic matrix in the form of an amorphous mixture of proteins and ice, and extracellular material, which, in the case of isolated cells will be the suspending medium, or, in the case of organised tissues, the intercellular material or 'ground substance'. Rough or random fracture faces of this type, would appear identical on replicas of both parts of the fractured material. The second type of fracture face revealed, is a relatively smooth *fissure* face carrying particles, and produced by a process of fission where the fracture follows the natural discontinuity or weakness associated with a frozen membrane, either plasma membrane, or the membrane of a cytoplasmic organelle. Appreciation of the nature of membrane-associated fracture faces and their relation to the plane of fracture is crucial to the interpretation of micrographs.

1 Plane of fracture of biological membranes. As the fracture traverses a frozen specimen, it will evidently approach individual membranes at different angles, and in some instances it will continue straight across them (cross-fracture) so that they will be seen 'edge-on' in replicas, i.e. just as they are seen in conventionally processed thin sections (Fig. 36). In other instances, the fracture deviates and runs parallel with the membrane for variable distances thus revealing extensive membrane-associated fracture faces.

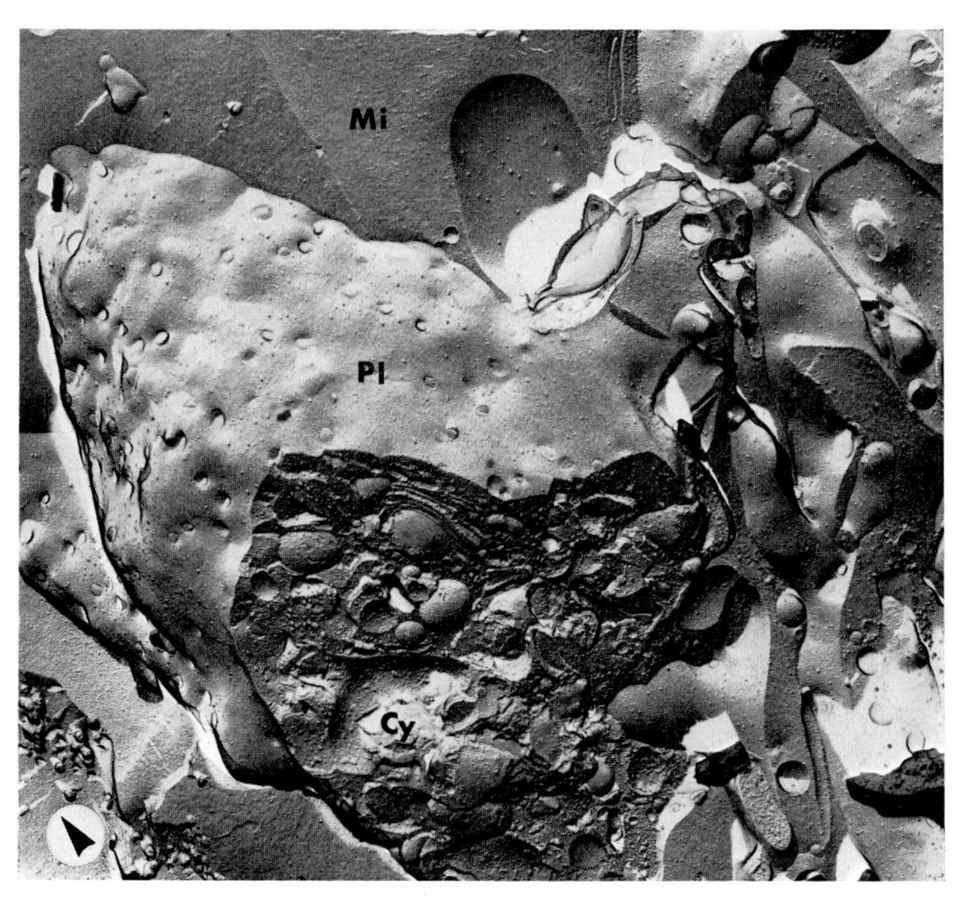

FIG. 35. Replica of cell from dermis of human foetal skin. Mi, is a fracture face of the extracellular milieu, and Cy a fracture face of the cytoplasm. Both of these are classed as 'random' fracture faces in contrast to the fracture face of the plasma membrane (Pl) which is classed as a 'fissure' face. × 26,000.

Uncertainty concerning the exact spatial position with respect to the membrane of the plane of fracture in the latter instance dates from the early work of Mühlethaler *et al.* (1965) and others (Weinstein and Bullivant, 1967; Weinstein, 1969). Their original conclusion was that extensive areas seen on their micrographs represented *en face* views of the true outer and inner surfaces of cell membranes. If this were a true interpretation, it would mean that the fractured tissue parts along two planes, (*a*) along an interface between plasma membrane and extracellular material, and (*b*) along an interface between plasma membrane and cytoplasm. Four different types of fracture face should be demonstrable in these circumstances, i.e. the true outer and inner surfaces of the membrane, and their respective complementary faces of extracellular material and cytoplasm (Fig. 37). However, in all instances,

FIG. 36. Replica of mature human epidermal cells. The nucleus of one cell (Nu 1) has been cross-fractured so as to reveal the nucleoplasm and an 'edge-on' view of the nuclear membrane (Me) as in thin sections. An *en face* view of the nuclear membrane of the second nucleus (Nu 2) has been revealed, and nuclear pores are clearly evident. Ap, line of cellular apposition. × 19,000.

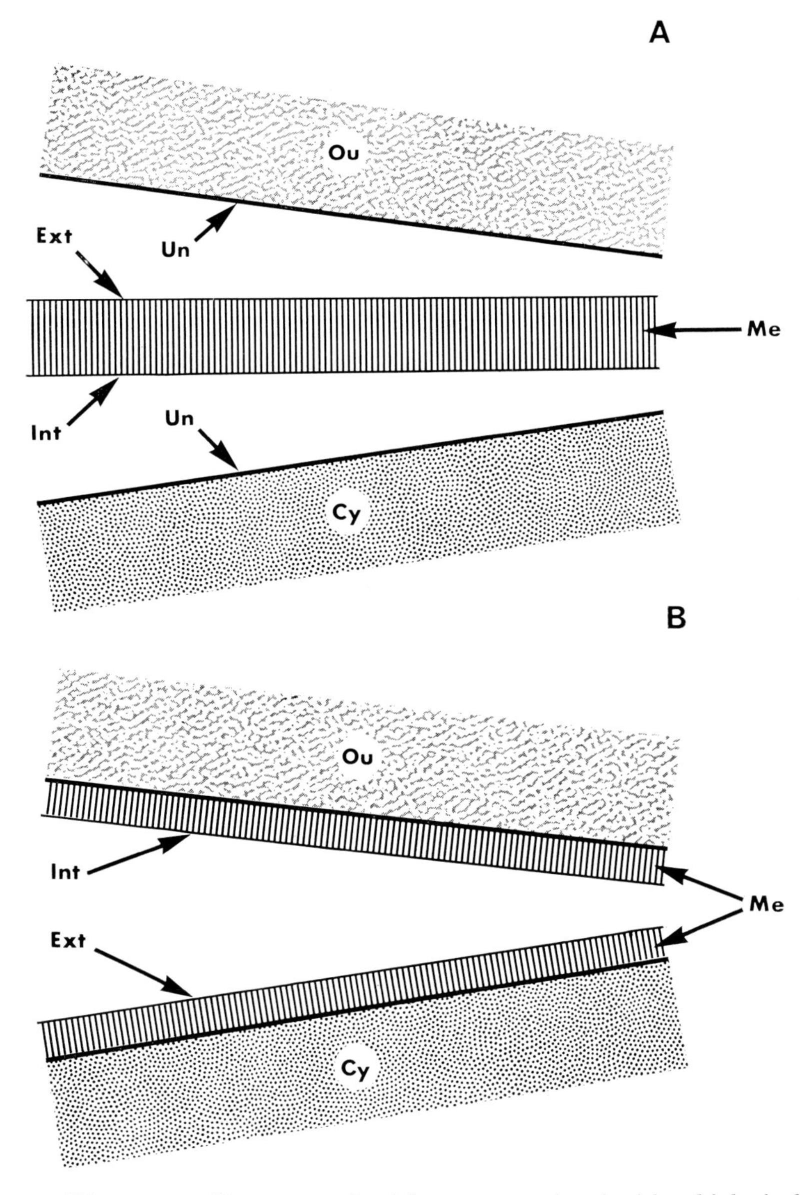

FIG. 37. Diagrams to illustrate path of fracture associated with a biological membrane. A. According to earlier interpretations. It was thought that the fracture could reveal true outer (Ext) and inner (Int) surfaces of the membrane (Me). However, theoretical complementary fractures faces (Un) of outer material (Ou) and of cytoplasm (Cy) were never identified. B. After Branton (1966). The two fracture faces (Ext. Int) are produced by 'splitting' of the membrane. This view now commands general acceptance, and applies to most membranes.

only two types of face, instead of four, were identifiable in micrographs of replicas. This clearly indicated that the fracture process associated with a frozen membrane follows only one plane.

Branton (1966; 1969) suggested that this plane lies within the hydrophobic interior of the membrane, which is therefore 'split' to reveal (on a single replica) either of two internal fracture faces (Fig. 37). Experiments involving deep sublimation of ice overlying membranes, combined with the attachment of markers to the true outer surface, tend to confirm this hypothesis (Tillack and Marchesi, 1970; Pinto da Silva and Branton, 1970; Branton, 1971; Pinto da Silva, Douglas and Branton, 1971). In these experiments, markers such as fibrous actin or ferritin, previously attached to the true outer surfaces were never seen on the original fracture faces of membranes. Deep sublimation of ice, however, revealed an additional face (etch-face) on which the markers were identified. This face, evidently therefore representing the true outer surface of the membrane, overlay the original fracture face, and was separated from it by a step which could represent a portion of the membrane thickness (Fig. 38). It seemed reasonable on this evidence to conclude that the fracture traverses the interior of the membrane, and that the faces revealed are generated within it. Evidently in this circumstance, one of the two faces will be directed towards the material internal to, or enclosed by the membrane, and the other towards the material external to it ('Int' and 'Ext' type faces respectively as defined in section on Nomenclature, p. 27).

The demonstration by Wehrli *et al.* (1970) that the plane of fracture associated with a membrane is unique, is in line with Branton's interpretation, and so, in the main, are the results of experiments involving production of thin sections at right angles to the plane of fracture of previously freeze-fractured material (Bullivant, 1969; Nanninga, 1971; Hereward and Northcote, 1972). In most of the latter experiments it was clear that the fracture traversed the interior of the membranes, but, occasionally, it appeared to have passed along the true surfaces. Notwithstanding the somewhat equivocal nature of these results, most workers nowadays operate on the basis that Branton's split-membrane hypothesis applies to the majority of biological membranes. The occasional instances in which true membrane surfaces are revealed by freeze-fracture may be thought to be due to particular processing steps prior to fracturing, or to specialised structural features

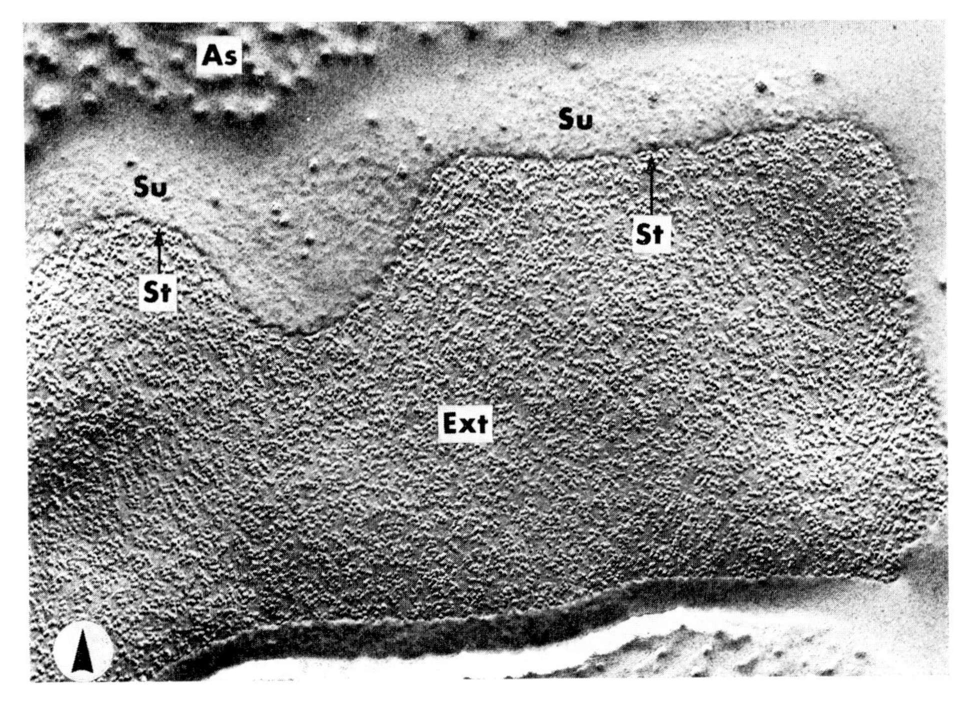

Fig. 38. Freeze-fractured and sublimated erythrocyte. On fracture, only the face Ext with many particles was revealed. Following sublimation the additional face, Su, was revealed between the Ext face and the lowered general ice level exhibiting asperites (As). The face Su is the true outer surface of the erythrocyte membrane which is never revealed on straightforward freeze-fracture, and the face Ext is an internal face revealed by splitting of the membrane. The step St between the two faces represents, therefore, a portion of the membrane thickness, and the particles on the Ext face must lie within the membrane. × 68,000.

of individual membranes and their immediate environments. Or, to factors at present unsuspected.

Whereas it may be accepted that the plane of fracture generally traverses the interior of membranes, and therefore reveals elements of their matrices, its exact spatial position with respect to membrane thickness and supposed structure remains uncertain at present. Before this question can be further pursued, it is necessary to consider in greater detail two features of the fracture faces associated with membranes, i.e. membrane-associated particles, and complementarity.

2 *Membrane-associated particles*. The most striking feature revealed on fracture faces of membranes are individual 80–100 nm particles

referred to as 'membrane-associated particles' (Figs. 39 and 41). Originally suspected of being artefacts of contamination, it now seems clear that they are real components of membranes, and, in that the plane of fracture lies within membranes, that they are situated in the membrane matrix (Flower, 1971). The possibility of their being extraneous contaminants was ruled out by Branton and Park's (1967)

Fig. 39. Membrane-associated particles (M.A.P.) on Ext fracture face of human erythrocyte plasma membrane. × 176,000.

observation that their numbers and dimensions remained unchanged when the time interval between fracturing and replication was extended to 10 minutes. That they represent integral membrane components is further suggested by observations on epidermal cells (Breathnach *et al.*, 1973). Particles are present on fracture faces of general plasma membranes of cells of all epidermal layers deep to the stratum corneum, but (apart from desmosomes) are virtually absent from membrane faces of cells of this latter layer (Fig. 40). It is generally accepted that the plasma membrane of the stratum corneum cell is structurally different

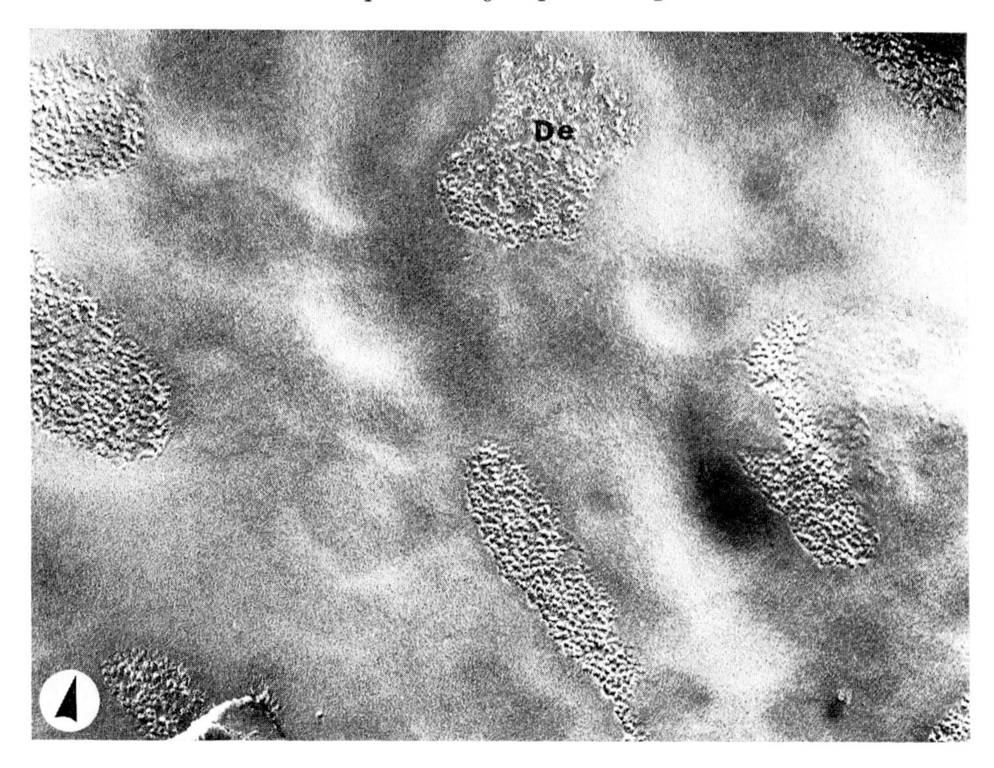

FIG. 40. Fracture face Ext of plasma membrane of cell of human stratum corneum. Note, apart from aggregated particles associated with desmosomes (De), complete absence of membrane-associated particles. × 30,000.

to membranes of cells of underlying layers, and the difference in particle numbers is very likely associated with this.

Evidence is accumulating that membrane-associated particles (M.A.P.) are proteins, or lipid-protein aggregates. Thus, fracture faces of membranes low in protein, such as nerve myelin, carry few, or no particles whereas faces of the erythrocyte membrane, which has a high protein content, have many particles (Fig. 39). Particles of the erythrocyte membrane are rendered mobile by changing the pH of solutions (Pinto da Silva, 1972) or by applying protein digesting enzymes, and these enzymes cause aggregation and final removal of the particles (Branton, 1971). Biochemical analysis of the same particles by Marchesi *et al.* (1972) led them to suggest that they form a part of glycoprotein molecules which have a specific orientation and serve as receptors or

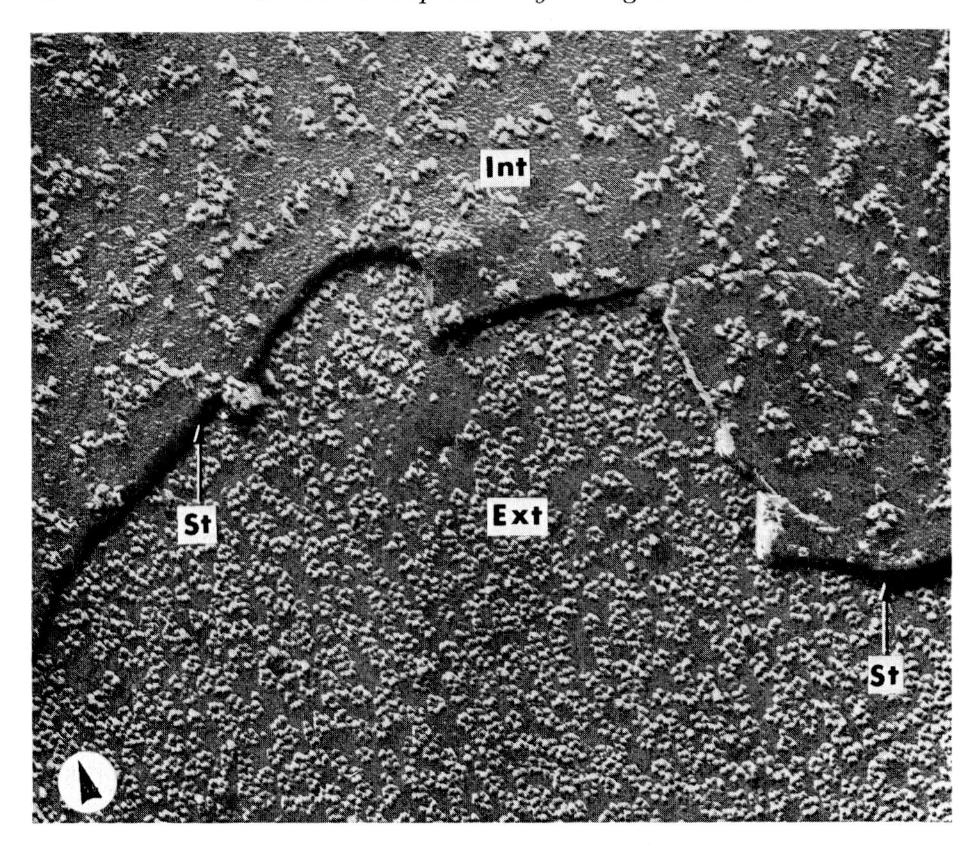

Fig. 41. Ext and Int faces of apposed plasma membranes of cells in choroid coat of guinea-pig eye. Note more numerous particles on the Ext face. St. step including intercellular material. × 88,000.

recognition sites for other molecules. It has further been suggested that individual particles may represent specific enzymes, and if this be so, the possible future combination of cytochemical and freeze-fracture techniques could aid in their more precise identification and location within particular membranes.

For most membranes, the concentration of particles is greater on the 'Ext' (A) face, than on the 'Int' (B) face (Fig. 41). As more and more tissues are being examined it is becoming clear also that there are considerable variations in numbers, grouping, and range of particle size, as between different membranes, and there is a tendency to relate these variations to the 'activity', physiological state or specialised function, of the membrane. However, until detailed particle size and particle density measurements, such as those carried out by Branton (1969) and by Staehelin (1973) are available for a variety of membranes of different tissues, these views remain somewhat speculative.

C *Complementarity of Membrane Fracture Faces*

If, as appears to be the case, the two faces revealed by freeze-fracture are produced by internal splitting of a membrane, then, there should be perfect complementarity or matching between the two, and fitting them together again should result in reconstitution of an intact membrane. However, where most membranes are concerned, even in instances where replicas taken from both sides of the fracture (matched replicas) are available for inspection, the great majority of workers report that an exact match does not result. This lack of perfect complementarity relates mainly to a discrepancy between the number of membrane-associated particles on one face, and depressions capable of accommodating them on the other face.

Staehelin (1973) claims to be able to demonstrate adequate numbers of depressions in most of his recent material, and suggests that their absence is indicative of a replica of suboptimal quality. However, the situation is probably not quite as simple as this. Even in the early days, some depressions were demonstrable (Branton, 1966), and readily so in certain specialised areas of membranes, such as the nexus, or gap junction (Chalcroft and Bullivant, 1970). In this situation (Fig. 54) depressions of diameter 3 ± 1 nm are apparent even in poor replicas, yet, on the immediately adjacent fracture face of the general plasma membrane where one might expect to find even larger depressions capable of accommodating M.A.P.s, none are apparent. A study of the literature in general will confirm that the presence or absence of depressions is associated less with quality of replica, than with the particular tissue being investigated. Whereas, it is possible that depressions may be obscured through filling up with excess of shadowing material or extraneous contaminants, it is extremely doubtful if this is the full explanation for the variations in complementarity observed. Branton (1973) draws attention to two other possibilities, related to the manner in which proteins may be intercalated into the membrane. If a protein is inserted with minimal intercalation, fracture could produce two nearly complementary replicas; if it completely traverses the membrane, it may become stretched and deformed plastically before fracturing (Meyer and Winkelmann, 1969; Bullivant *et al.*, 1972) and this could lead to lack of complementarity. Another possibility which should be considered is, that there may be variable loss of membrane matrix material during the process of fracturing and evaporation, i.e. material which would contain depressions capable of accommodating particles. Clearly, the fracture on entering the frozen membrane complex, follows the path of least resistance, and this could coincide with poorly attached material, possibly pure phospholipid, which does not contain water in its interior, and which, when frozen, does not form a continuum with water-containing components in contact with it. Loss of material could be attributed to the possibility that at very low temperatures and under vacuum, forces produced during the fracturing process may be sufficient to disperse it. Also, the evaporation process itself, which subjects the revealed surfaces to heavy metal bombardment, might effectively dislodge loosely attached material. Particles remaining on the fracture faces could represent active elements containing water and firmly attached at particular sites. If the proposed poorly attached material were to be more firmly bound, as for example, following chemical fixation, an alteration in the path of the fracture might be expected. This, in fact, is the case following fixation with osmium tetroxide (Meyer and Winkelmann, 1970; James and Branton, 1971).

One might conclude from the above discussion that variations in detailed complementarity between membrane fracture faces, far from indicating poor technique, may, in fact express underlying chemical and molecular configurations, and therefore, structural and functional differences between membranes, or between different areas of the same membrane. The fact that such variations can occur, makes it difficult to specify the exact plane of fracture in terms of proposed theories of basic membrane structure, and this in turn, makes it difficult to deduce the exact spatial and relative positions of the structures revealed. However, it may be said that to date, the freeze-fracture technique has greatly advanced knowledge of membrane ultrastructure, and the inherent precision of the method augurs well for further advances.

III ANALYSIS OF TYPICAL REPLICAS

One can now proceed to the analysis of individual micrographs of replicas in accordance with the principles outlined above. As a simple example, one may choose a replica of a fractured erythrocyte suspended in plasma mixed with buffered glycerol medium (Fig. 42A). The first step is to determine the direction of platinum shadowing. Remembering that the convention is to print negatives so that shadows appear white on micrographs, one notices an extensive area (Sh) with apex pointing upwards and towards the right. This indicates the general direction from which shadowing platinum material was evaporated, to be from below and left. This is confirmed, and can be more precisely defined (arrow at bottom left) by noting the direction of shadows cast by the organelle (Or) and the small particle (p). In order to define the main topographical features of a replica, it is useful for the beginner to consider what the fracture has revealed as one proceeds along an arbitrary line, such as A–B on the micrograph, and to build up in his mind (or on a simple diagram) a profile of the shadowed surface and of the fractured tissue, such as that presented in Fig. 42B. Proceeding along the line from A therefore, the fracture can be seen to reveal in succession: 1. a rough fracture face of the suspending medium (Mi); 2. a fracture face of the erythrocyte membrane (Ext) which is directed towards the exterior of the cell, i.e. towards the medium; the heavier deposit of platinum on this fracture face indicates that it is 'stepped-up' in relation to the adjacent fracture face of the medium, as indicated on the profile; 3. a face (Hb) resulting from fracture of haemoglobin contained by the membrane; as the density of platinum deposit on this fracture face is approximately equal to that of the medium face, it can be represented on the profile as lying parallel with the latter; 4. a fracture face of the erythrocyte membrane (Int) directed towards the interior of the cell as evidenced by the fact that it is overlaid by the haemoglobin; consideration of the density of platinum deposited on this face, in turn, indicates that it is stepped-up obliquely in relation to the haemoglobin, as indicated on the profile; 5. a fracture face of the medium beyond the cell. Having established the nature of the various faces, and their relation to each other, each can now be examined in detail. In this instance, the only points of note relate to the higher

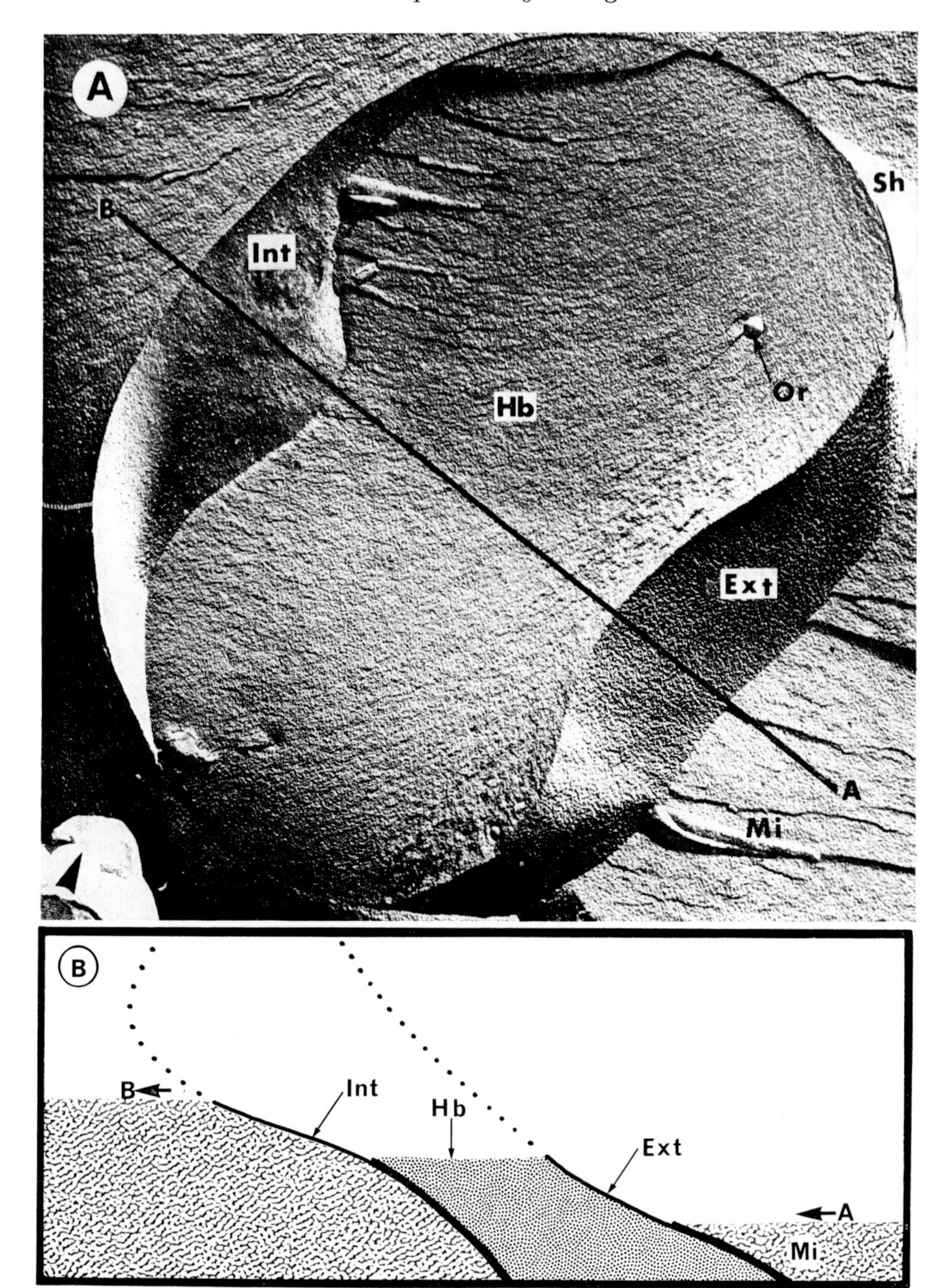
A
Sh
B
Int
Hb
Or
Ext
A
Mi
B
B
Int
Hb
Ext
A
Mi

concentration of M.A.P.s on the 'Ext' face as compared with the 'Int' face, and the presence of the organelle (which can be described as a convexity by considering the direction of shadowing) within the haemoglobin. This is probably a 'Heinz body'.

A more complicated replica is presented in Fig. 43A, but in analysing it, the same principles apply. The direction of shadowing, as indicated, can be determined by noting shadows related to, or cast by spherical organelles and filaments present in the cytoplasm of the cells. Proceeding along the line A–B, the fracture can be seen to have revealed in succession; 1. a random fracture face which could be either cytoplasm (Cy) or intercellular material (Ic) and which is, in fact the latter because it is limited towards the left by a scarcely visible cross-fractured membrane (Mec); 2. a membrane fracture face, which, considering the direction of shadowing and the density and distribution of the platinum deposit is evidently elevated with respect to the intercellular material (see profile, Fig. 43B); 3. a narrow zone of minimal density of platinum deposit which is followed by 4. another membrane fracture face. Obviously, here we are dealing with the apposition between two cells, with the 'Ext' face of one plasma membrane separated from the 'Int' face of an apposed plasma membrane by an intercellular step (Ic). Next, 5. follows cellular cytoplasm containing 'concave' and 'convex' spherical organelles, as well as filaments, and finally, 6. a fracture face of the plasma membrane of the cell directed towards its exterior (Ext). All of these features are represented in the profile in Fig. 43B.

In Fig. 44 a replica is presented for analysis by the reader. Examine it first of all from the point of view of quality, in terms of evidence of artifact, contamination, etc. Then determine the direction of shadowing, and construct a profile of the fracture face along the line A–B as in Figs. 42 and 43.

Matched replicas (Figs. 45 and 46) are analysed on exactly similar lines, corresponding areas on each micrograph being usually readily apparent.

FIG. 42. A, replica of human erythrocyte with fracture faces Ext and Int associated with the plasma membrane and limiting the cell contents, Hb. The biconcave disc shape of the erythrocyte is clearly recognisable. × 22,000. B, profile along the line A–B. The overall shape of the erythrocyte is indicated by representing the portion fractured away in dotted outline, and by relating the portion remaining to the fracture faces revealed. See text for analysis.

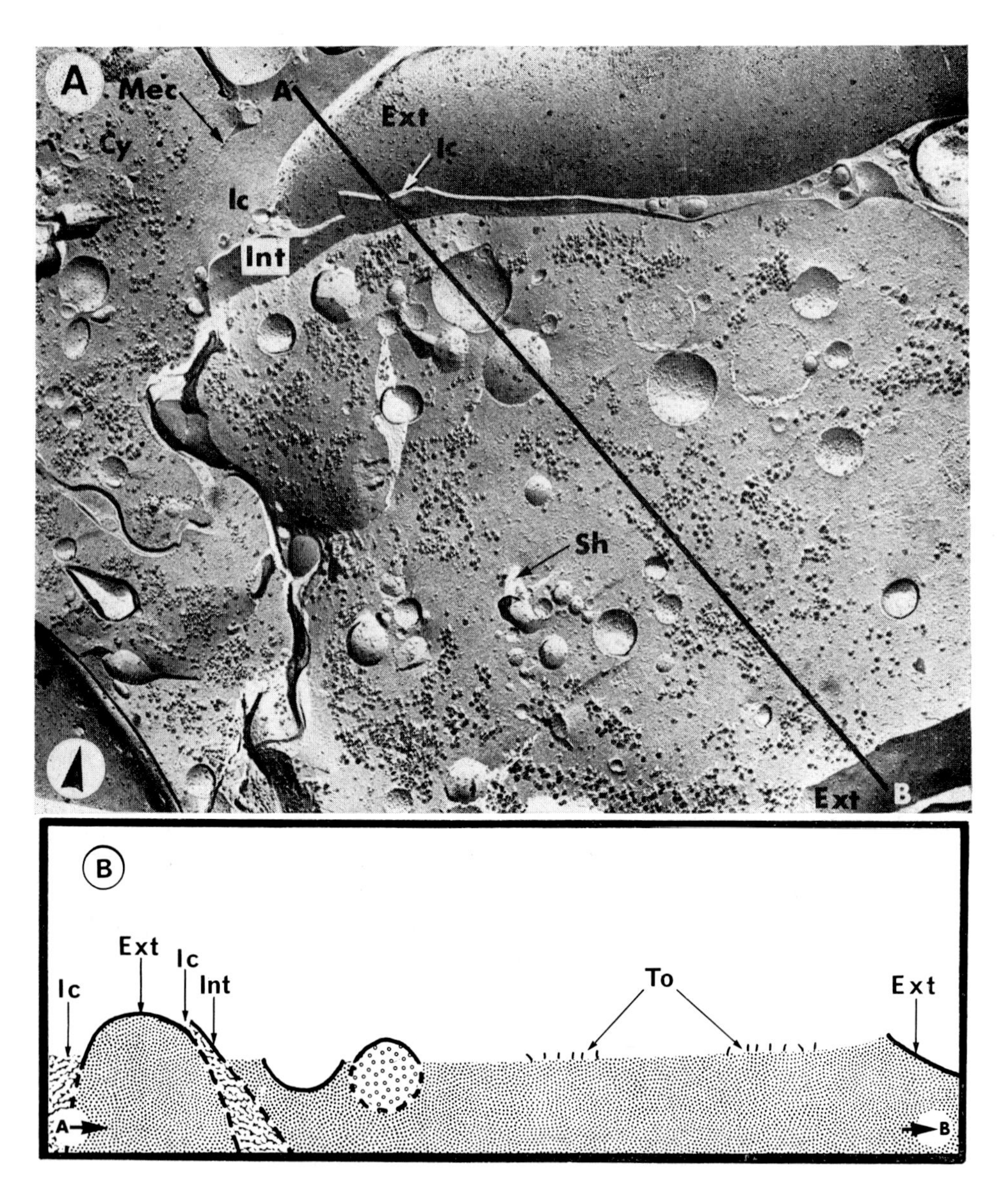

FIG. 43. A, replica from human foetal epidermis which is analysed in the text and in the diagram Fig. 43B. Sh, shadow cast by cytoplasmic organelle. × 28,000. B, reconstruction of features revealed by fracture along the line A–B in Fig. 43A.

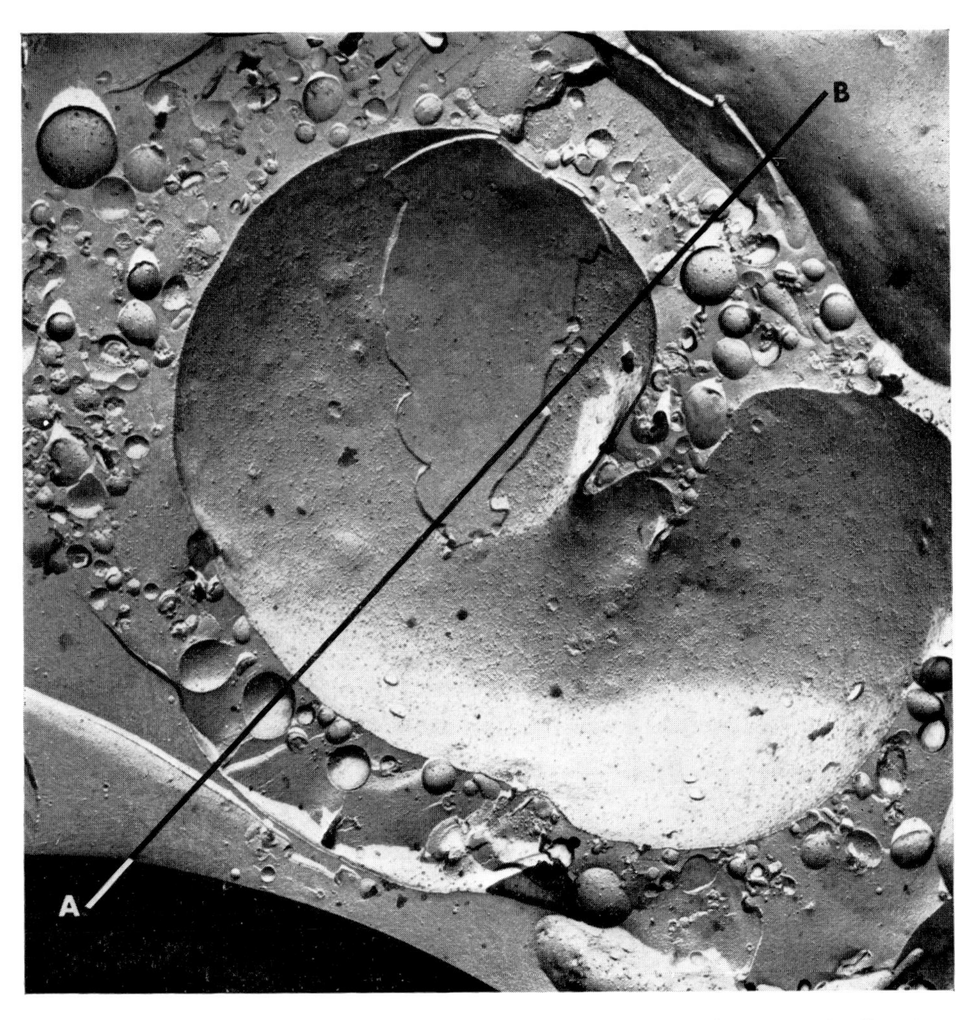

FIG. 44. Replica for analysis by the reader. Construct a profile along the line A–B as in Fig. 43B. × 22,400.

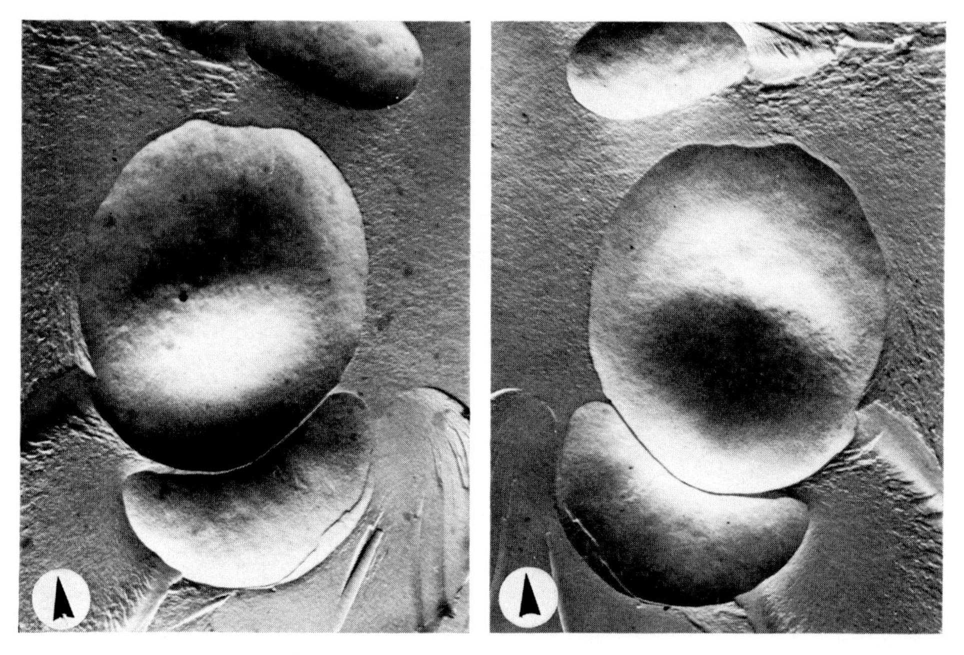

FIG. 45. Matched replicas of erythrocyte suspended in buffered glycerinated plasma. × 11,000.

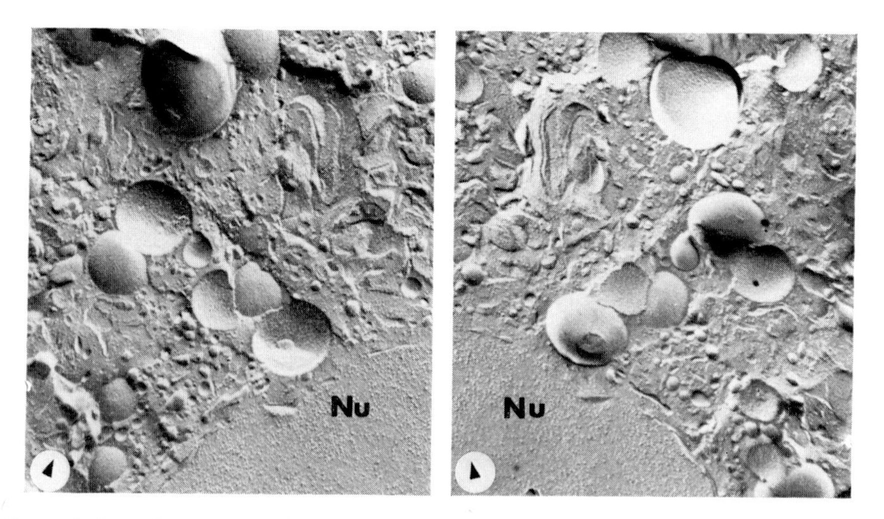

FIG. 46. Matched replica from mouse jejeunal epithelium. Nu, nucleus. × 18,000.

Obviously, with material which has not been examined previously, it is a great help to have corresponding micrographs of stained thin sections for comparison with replicas. Since many workers' conception of the ultrastructure of tissues is based upon the two-dimensional appearance presented by the former, some find it a bit difficult, at first, to equate the two. In this connection the construction of simple profiles or models is a great help in making a synthesis. Use of a goniometer stage which allows examination of the replica at varying angles of tilt, and the production of stereo-pairs of micrographs, can often yield valuable extra information (Steere, 1973). The micrograph is the final product of the whole technique, and its quality, and the information it can provide depends upon the various factors already discussed in Chapters 2 and 3.

REFERENCES

Branton, D. (1966). *Proc. Nat. Acad. Sci.* **55**, 1043–1056.
Branton, D. (1969). *Ann. Rev. Plant Physiol.* **20**, 209–238.
Branton, D. (1971). *Phil. Trans. Roy. Soc.* **B261**, 133–138.
Branton, D. (1973). *In* 'Freeze-etching, techniques and applications' (E. L. Benedetti and P. Favard, eds) 107–112. Société Française de Microscopie Électronique, Paris.
Branton, D. and Park, R. B. (1967). *J. Ultrastruct. Res.* **19**, 283–303.
Breathnach, A. S., Goodman, T., Stolinski, C. and Gross, M. (1973). *J. Anat. Lond.* **114**, 65–81.
Bullivant, S. (1969). *Micron* **1**, 46–51.
Bullivant, S., Rayns, D. G., Bertaud, W. S., Chalcroft, J. P. and Grayston, G. F. (1972). *J. Cell. Biol.* **55**, 520–524.
Calbick, C. J. (1951). *Bell Syst. Tech. J.* **30**, 798–824.
Chalcroft, J. P. and Bullivant, S. (1970). *J. Cell. Biol.* **47**, 49–60.
Flower, N. E. (1971). *J. Ultrastruct. Res.* **37**, 259–268.
Hereward, F. V. and Northcote, D. H. (1972). *J. Cell Sci.* **10**, 555–561.
James, R. and Branton, D. (1971). *Biochem. Biophys. Acta* **233**, 504–512.
Marchesi, V. T., Tillack, T. W., Jackson, R. L., Segrest, J. P. and Scott, R. E. (1972). *Proc. Nat. Acad. Sci.* **69**, 1445–1449.
Meyer, H. W. and Winkelmann, H. (1969). *Protoplasma* **68**, 253–270.
Meyer, H. W. and Winkelmann, H. (1970). *Protoplasma* **70**, 233–266.
Mühlethaler, K., Moor, H. and Szarkovsky, J. W. (1965). *Planta (Berl.)* **67**, 305–323.
Nanninga, N. (1971). *J. Cell Biol.* **49**, 564–570.

Pinto da Silva, P. (1972). *J. Cell Biol.* **53**, 777–787.
Pinto da Silva, P. and Branton, D. (1970). *J. Cell Biol.* **45**, 598–605.
Pinto da Silva, P., Douglas, S. D. and Branton, D. (1971). *Nature* **232**, 194–196.
Shaefer, V. J. and Harker, D. (1942). *J. App. Phys.* **13**, 427–433.
Staehelin, A. S. (1973). *In* 'Freeze-etching, techniques and applications' (E. L. Benedetti and P. Favard, eds) 113–134. Société Française de Microscopie Électronique, Paris.
Steere, R. L. (1973). *In* 'Freeze-etching, techniques and applications' (E. L. Benedetti and P. Favard, eds) 223–255. Société Française de Microscopie Électronique, Paris.
Tillack, T. W. and Marchesi, W. T. (1970). *J. Cell Biol.* **45**, 649–653.
Wehrli, E., Mühlethaler, K. and Moor, H. (1970). *Exptl. Cell Res.* **59**, 336–339.
Weinstein, R. S. (1969). *In* 'Red Cell Membrane Structure and Function' (G. A. Jamieson and T. J. Greenwald, eds) 36–82. Y. B. Lippincott, Philadelphia.
Weinstein, R. S. and Bullivant, S. (1967). *Blood* **29**, 780–789.

5

Typical Appearance of Cell Components in Replicas

In this chapter the appearance of the commoner cell components and organelles will be described in order to provide a general background for the interpretation of replicas of individual cell types and organised tissues, as presented in the chapter following.

I PLASMA MEMBRANE

A General Features

As already explained (Chapter 4) freeze-fracture of a biological membrane frequently reveals two complementary internal fracture faces, and where the plasma membrane is concerned, these are directed respectively towards the interior ('Int' face) or the exterior ('Ext' face) of the cell. When individual isolated cells appear fractured in their entirety (Figs. 42 and 47) both faces of the same membrane may be revealed, a different face at opposite sides of the cell. With organised tissues consisting of closely apposed cells, e.g. epithelia, the two types of face, one or other associated with the plasma membrane of each of two different cells are seen, separated by a step which includes the intercellular interval or material (Figs. 41 and 48). Both faces carry membrane-associated particles (M.A.P.), and in general, these are more numerous on the 'Ext' face. Apart from questions as to the basic nature and significance of these particles (see Chapter 4), others, relating to variations in their size, numbers, distribution, and arrangement arise. It is becoming increasingly evident that these features may vary with species, tissue, phase of cell cycle, and physiological state of the cell. For example as already noted (Fig. 40), M.A.P. are almost completely absent from fracture faces of the general plasma membrane of

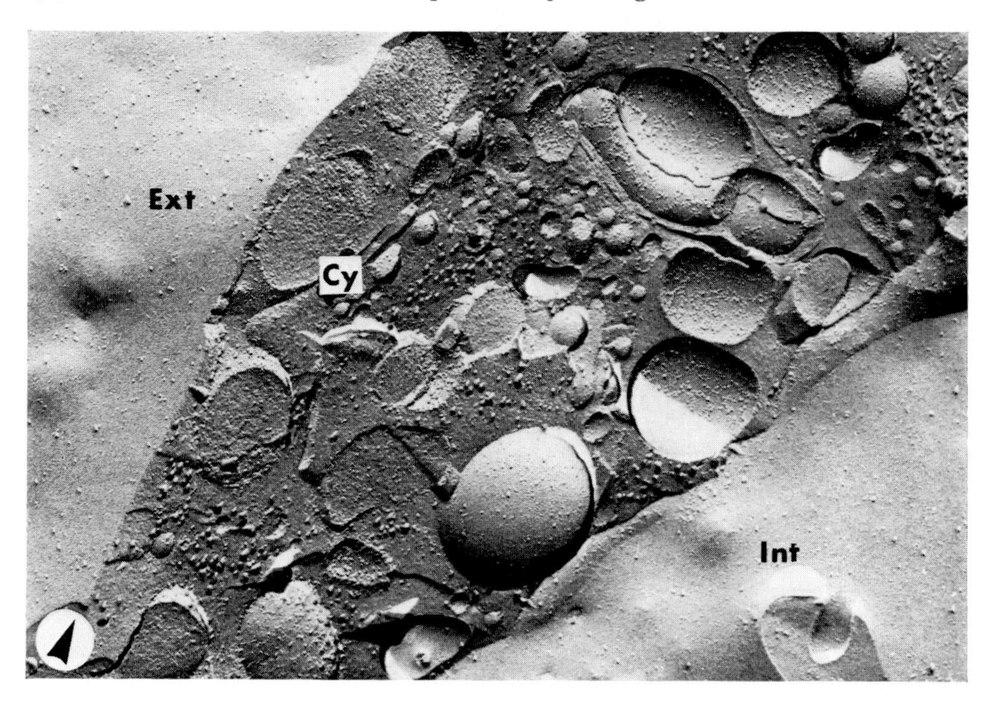

Fig. 47. Replica of cell from dermis of human foetal skin. The fracture has revealed two fracture faces of the plasma membrane, one (Ext) directed towards the exterior of the cell, and another (Int) directed towards the interior i.e. the cytoplasm (Cy). × 44,000.

the stratum corneum cell of human epidermis (Breathnach *et al.*, 1973). It seems likely that some of these features may also be affected in general or specific disease conditions.

B *Specialised Junctional Complexes*

Specialised areas of junctions between cells, exhibiting a variety of regular patterns have been extensively studied during the past decade. Farquhar and Palade (1963) divided these junctional complexes, as seen in thin sections, into the following three types: (1) *macula adhaerens* (desmosome), (2) *zonula adhaerens* (gap junction), (3) *zonula occludens* (tight junction). A fourth type, the *septate junction*, occurring frequently in invertebrates, is also recognised. With the advent of the freeze-fracture technique, a new and extended insight into the structure of these complexes has been achieved (McNutt and Weinstein, 1973).

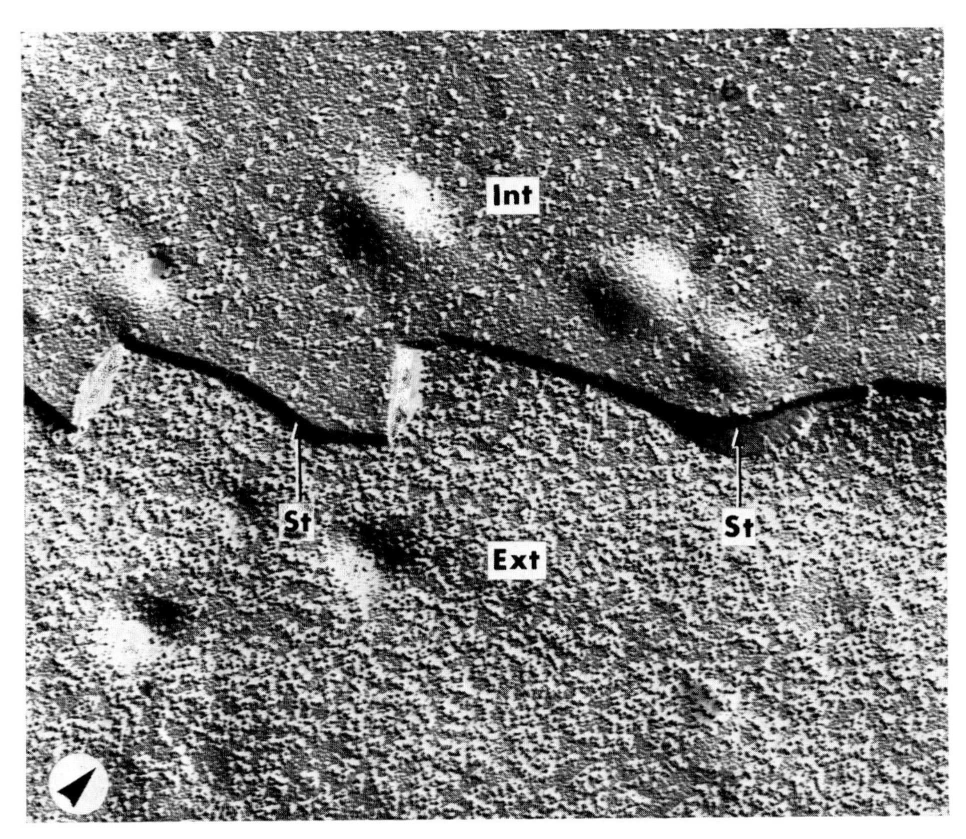

Fig. 48. Fracture faces of apposed plasma membranes of cells from choroid coat of eye of guinea-pig. A fracture face of one plasma membrane (Ext) directed towards the exterior of the cell, is separated by a step (St), which includes the intercellular material, from a fracture face of the plasma membrane of the apposed cell (Int) which is directed towards its interior. × 70,000.

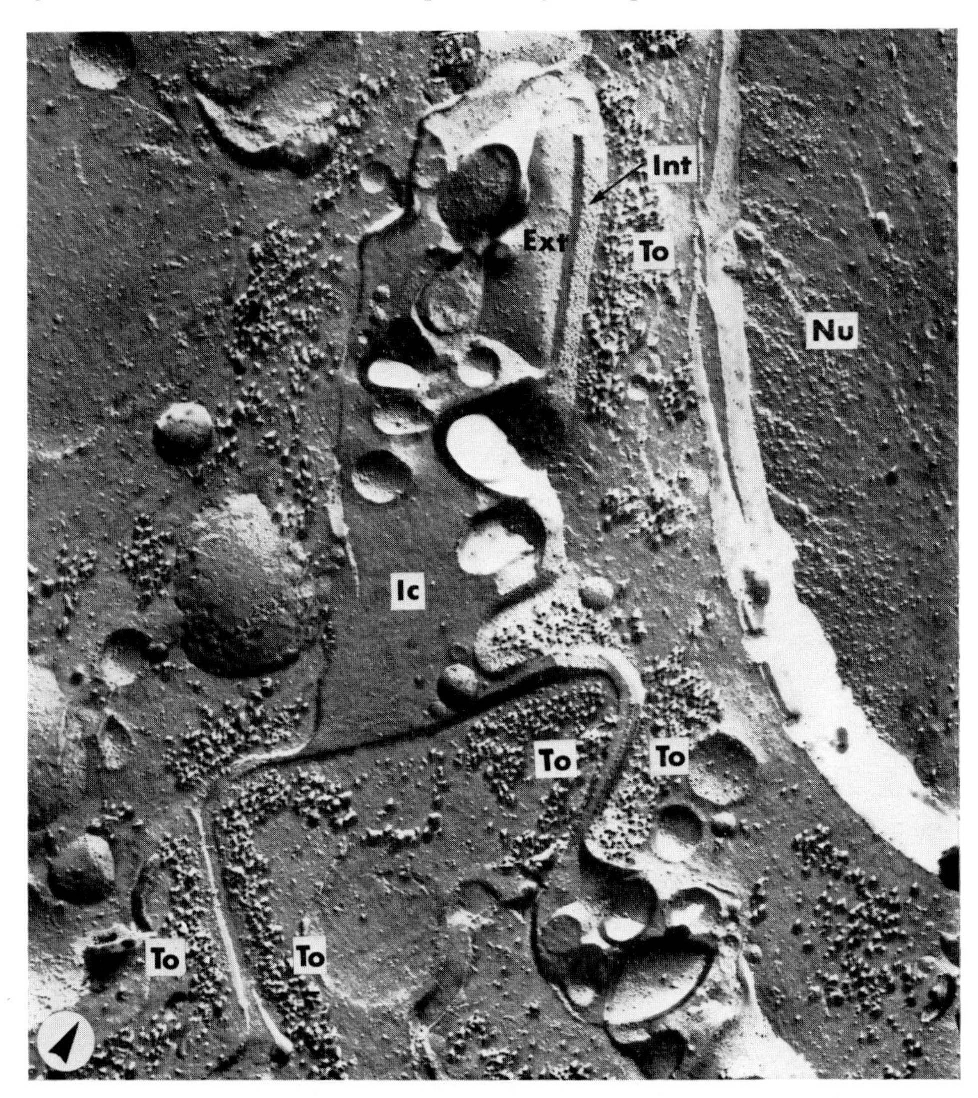

Fig. 49. Replica from epidermis of mature human skin. At the bottom of the field two cross-fractured desmosomes are seen with characteristic concentrations of tonofilaments (To) within the cytoplasm of the apposed cells. Towards the upper part of the field, another desmosome has been fractured so as to reveal Ext and Int faces of the plasma membranes involved. It is just apparent that there is a concentration of particles on the Int face (see also Figs. 50 and 51). Ic, intercellular space; Nu, nucleus. × 43,000.

1 Macula adhaerens (desmosome). When seen on cross-fracture (Fig. 49) the desmosome appears as a short zone where the plasma membranes of the apposed cells are non-undulent and run parallel with one another, about 30 nm apart. Within the cytoplasm of each cell a concentration of fractured tono-filaments, co-extensive with the desmosome is seen, and some of these filaments lie very close to the plasma membrane, i.e. within the territory occupied by the desmosomal plaque of stained thin sections. This latter structure is not identifiable in any positive sense in replicas. Likewise, in non-sublimated material, the intercellular intermediate dense line is not evident, but Staehelin *et al.* (1969) have reported some indication of it in sublimated material.

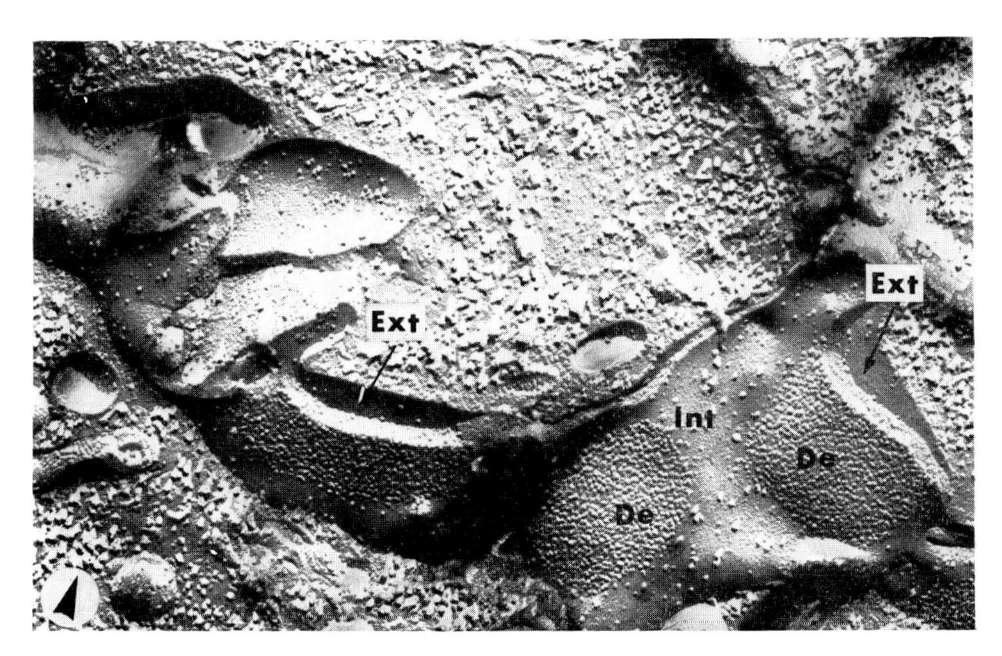

FIG. 50. Apposition of epidermal cells of mature human skin. Ext and Int faces of the apposed plasma membranes are revealed and aggregated desmosomal particles (De) are present only on the Int face. × 62,000.

En face views of fractured desmosomes (Figs. 50 and 51) reveal aggregations of closely packed particles on the 'Int' face of the plasma membrane, but only the occasional particle on the complementary 'Ext' face (Breathnach *et al.*, 1972). The average concentration of desmosomal particles, in human epidermis, is estimated as 700–800 per

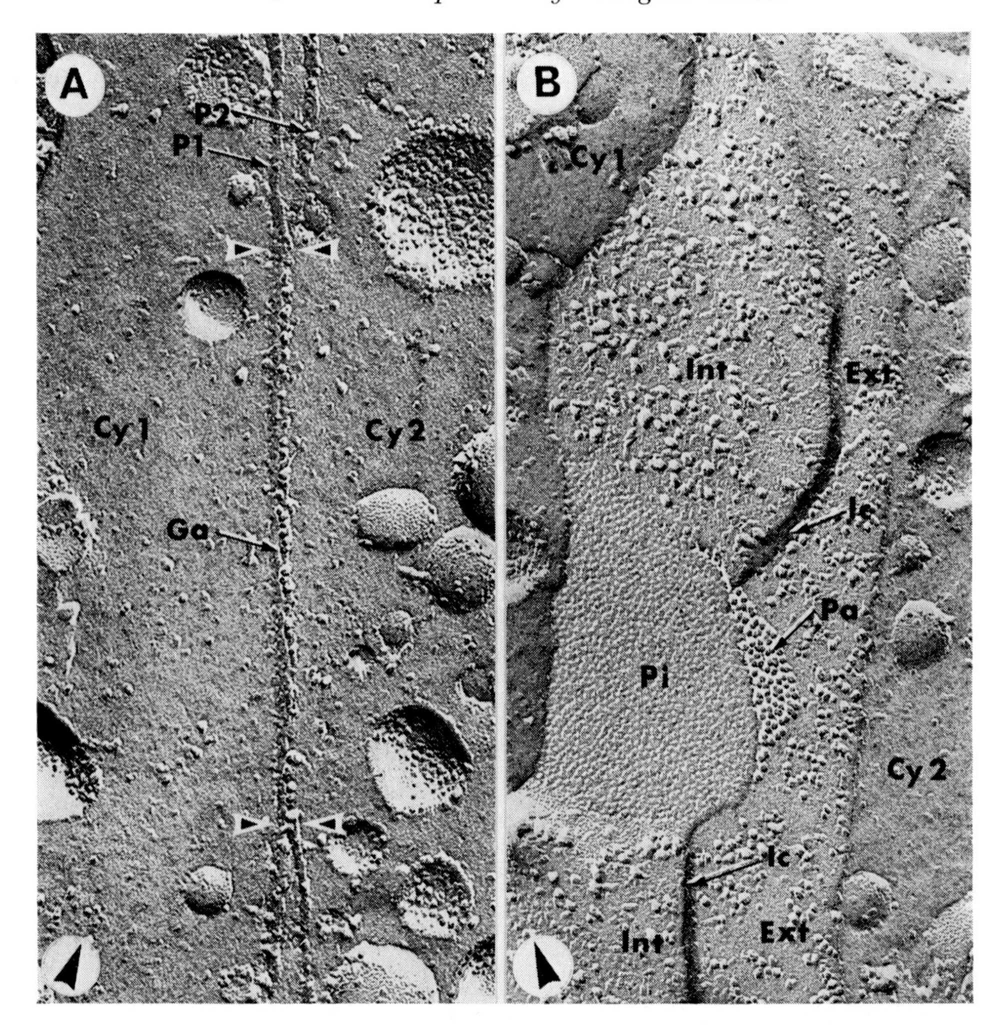

FIG. 53. Gap junction; cells of mouse liver. A, the gap junction on cross-fracture. Two cells with cytoplasm (Cy 1, Cy 2) and plasma membranes (P 1, P 2) are seen. In the region, Ga, between the upper and lower pairs of large arrows, it is clear that the plasma membranes of the two cells have become more closely apposed with great narrowing of intercellular material; this is a gap junction. B, *en face* view of gap junction. Here, Ext and Int fracture faces of the apposed plasma membranes are revealed. Note at the gap junction, presence of pits (Pi) within the substance of the Int face, and of closely aggregated particles (Pa) on the Ext face. Ic, intercellular material. × 154,500.

Weinstein, 1970; Chalcroft and Bullivant, 1970; Spycher, 1970; Friend and Gilula, 1972; Albertini and Anderson, 1974). At the gap junction, an area of uniformly distributed pits is revealed on fracture face 'Int', and a close-packed array of particles is present on the complementary fracture face, 'Ext' (Figs. 53 and 54). The pits on fracture face 'Int' have a diameter of 3 ±0·5 nm, and are distributed at random, or in a regular hexagonal lattice. In the latter instance, the pattern exhibits the usual 60° angle between the lines on which the pits are equidistantly aligned. The distance between the centres of the pits is a reasonably regular 8 ±1 nm (Fig. 54). The pits are within the general plane of fracture face 'Int', on which desmosomal and membrane-associated particles rest, and this is evident from the absence of any step delimiting the area of pits.

The particles revealed on fracture face 'Ext' of the plasma membrane at gap junctions are likewise spaced 8 ±1 nm centre to centre, and

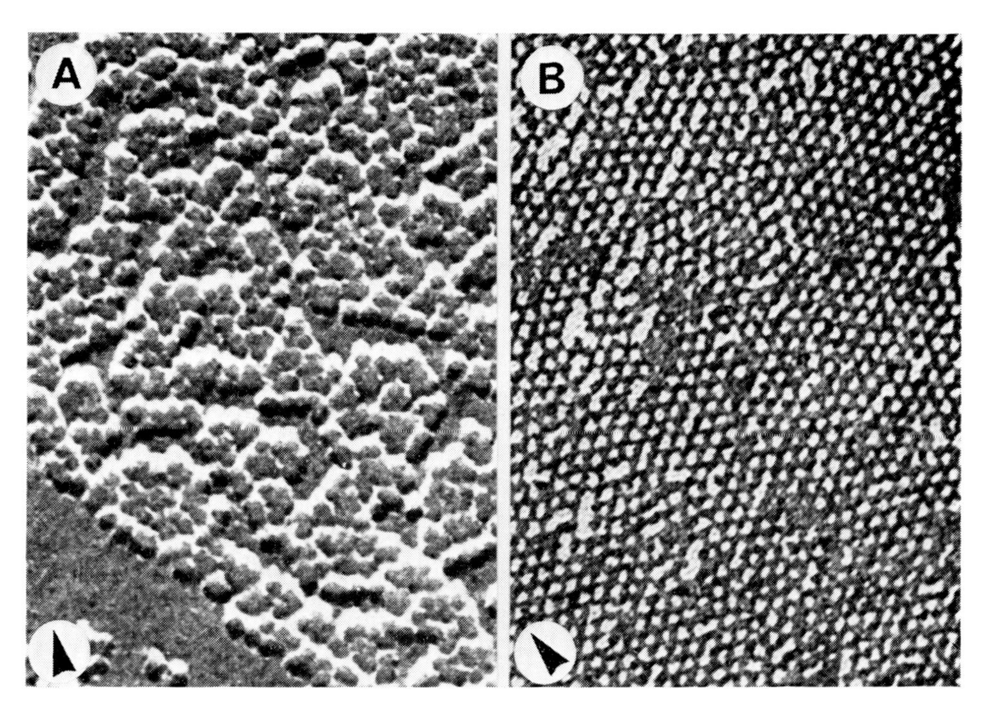

FIG. 54. Gap junction. A, aggregated particles on Ext face of plasma membrane; B, Pits on Int face of plasma membrane. × 282,000.

FIG. 55. Tight junction; epithelial cells of mouse jejeunum. At the junction, rows of particles (Pa) occupy furrows on the Int face of the plasma membrane, and on the complementary Ext face are ridges (Ri) carrying fewer particles. × 97,200. *See also* Fig. 56.

arranged randomly or in hexagonal lattice. Counts reveal an equal number of pits and particles per unit area of appropriate fracture face, and frequently, lines of pits are in direct alignment with lines of particles across the step delineating the two faces. Chalcroft and Bullivant (1970) have concluded from examination of matched replicas of both sides of the fracture that at gap junctions, there must be two layers of particles situated at different levels, and separated by a layer which has depressions on each side. They also refer to a layer of material 'sandwiched between the two arrays of particles'. In some micrographs, it is evident that this material is in direct continuity with the inter-cellular material, and it occupies a central position within the complex.

3 Zonula occludens (tight junction). The appearance of the tight junction depends to a considerable extent upon whether or not the tissue was chemically fixed before freezing (Staehelin, 1973). With gluteraldehyde fixed material (Chalcroft and Bullivant, 1970; Friend and Gilula, 1972) it is revealed as a linear, often branched ridge on fracture face 'Ext' of the plasma membrane with a complementary furrow on fracture face 'Int'. With unfixed material, the junctions appear as linear arrays of particles spaced approximately 10–13 nm on both fracture faces (Fig. 55). The particles are usually more numerous on fracture face 'Int', where they lie in furrows, than on fracture face 'Ext', where they lie on top of ridges. Gaps in the arrays of particles on both faces indicate that, on fracture, they are shared between the two faces, the majority remaining on fracture face 'Int'. In some micrographs, the step which reveals the intercellular space is seen to include the junctional particle, and the fracture planes associated with the two opposing plasma membranes may be seen to converge at a particle, with complete elimination of the intercellular material.

In some tissues (e.g. liver, intestine), the particles are disposed on the fracture faces in undulating lines, frequently forming closed loops, or T-configurations by joining adjacent arrays (Fig. 56). In others (e.g. epidermis) a linearly arranged tight junction may be seen surrounding a desmosome or a gap junction with a closed loop of particles.

The septate junction of the invertebrate epithelial cell bears a

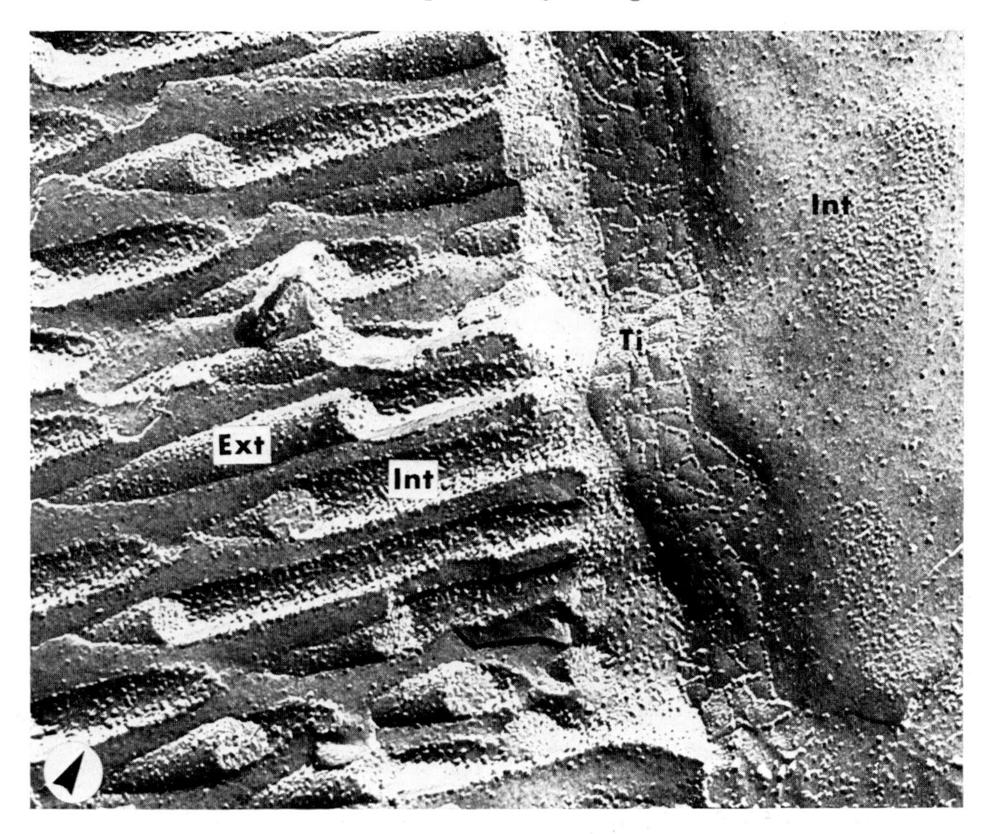

FIG. 56. Microvilli from brush border of epithelial cell of mouse jejeunum. Ext and Int faces of the membrane forming the villi are revealed. Note zone of tight junction particles (Ti) on fracture face Int of plasma membrane at bases of microvilli. × 73,000.

superficial resemblance to the tight junction, in that it is characterised by linear arrays of particles on the 'Ext' face of the plasma membrane, and ill-defined furrows or depressions on the complementary 'Int' face (Flower, 1971).

Many of the investigators cited above have presented three-dimensional interpretations of the arrangement of the various junctional complexes, based upon appearances in replicas, but at the moment, there is some difference of opinion concerning their exact organisation. This is partly due to difficulty in establishing the path of the fracture in relation to the general plasma membrane as well as to the more specialised regions involved in individual complexes.

C Surface Differentiations

1 Microvilli. These slender processes of the plasma membrane are characteristic of luminal borders of absorptive and other epithelia, and, as might be expected, they may be fractured so as to reveal either an 'Ext'- or 'Int'-type fracture face with characteristics similar to those of the respective faces of general plasma membrane (Fig. 56).

2 Pinocytotic vesicles. These are revealed as irregularly distributed 'bumps' or 'hollows' (Fig. 57) on the appropriate fracture face of the membrane ('Ext' or 'Int'), and their numbers give some indication of the level of micropinocytotic activity.

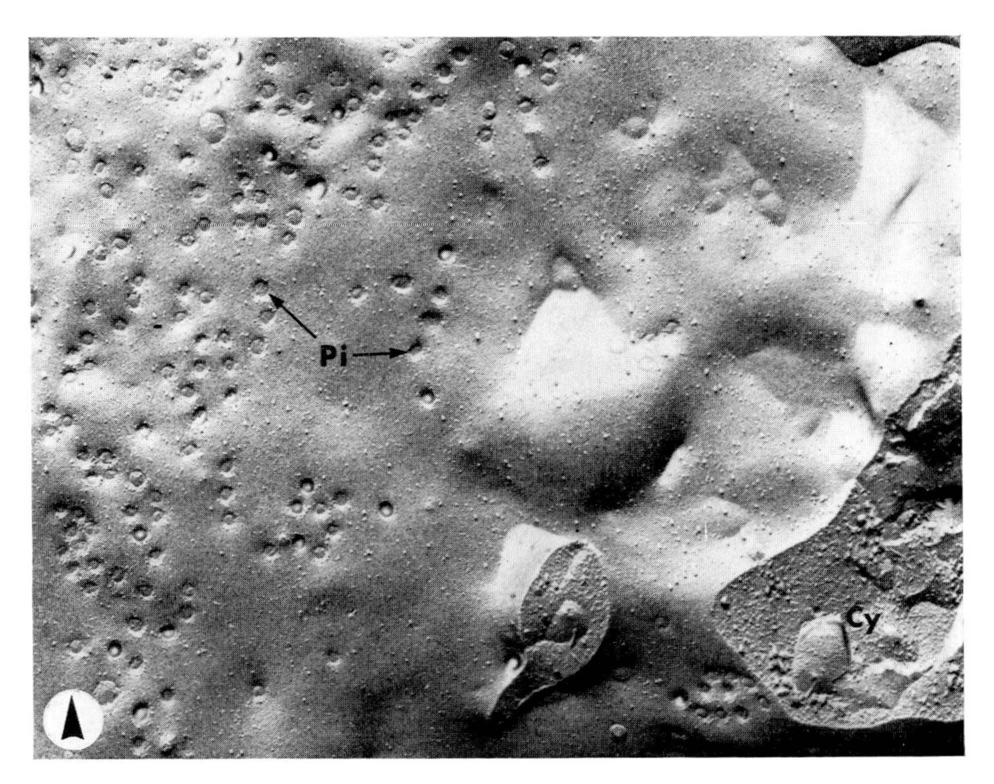

FIG. 57. Pinocytotic vesicles (Pi) on Ext face of plasma membrane of cell in dermis of human foetal skin. Cy, cytoplasm of cell. × 39,000.

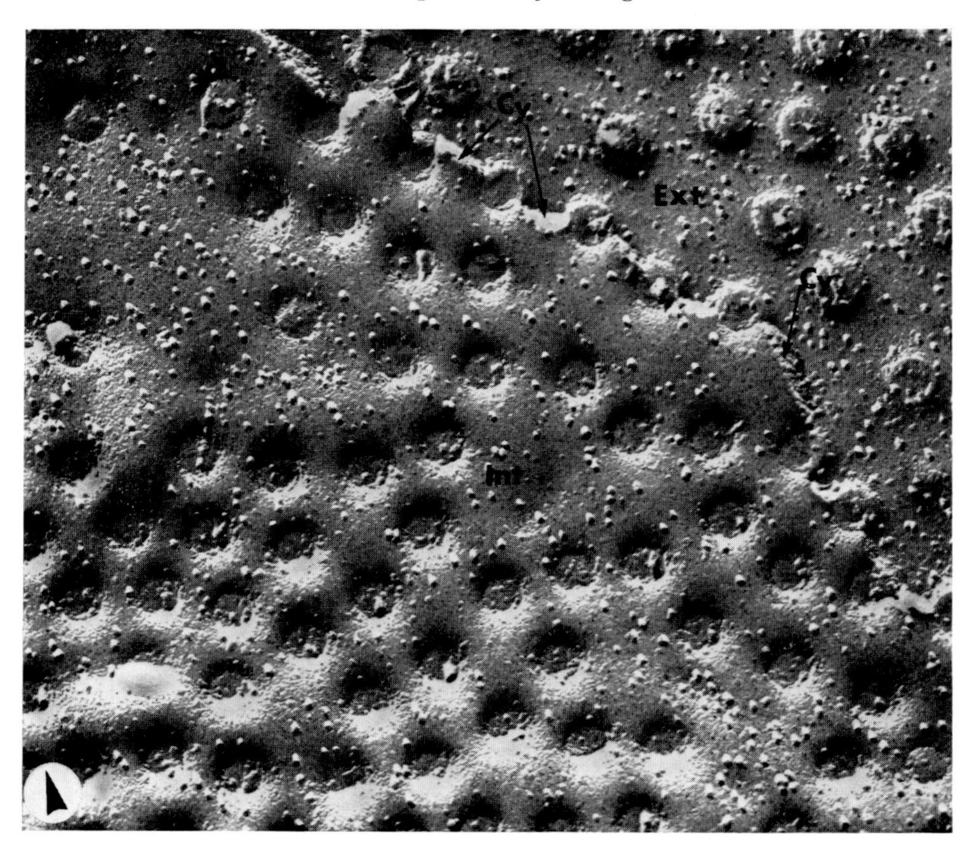

Fig. 58. Plasma membrane fenestrae appearing as regularly spaced circular depressions or elevations on Ext and Int faces respectively of plasma membrane of capillary endothelial cell from choroid coat of guinea-pig eye. Cy, cytoplasm of endothelial cell between the two faces × 107,000.

3 Plasma membrane pores (fenestrae). Extremely flattened vascular endothelial cells in certain situations exhibit pores or fenestrae in the plasma membrane, and these may be confused with pinocytotic vesicles on *en face* views of the membrane (Simionescu *et al.*, 1974). Like pinocytotic vesicles, fenestrae are revealed as circular elevations either cratered or flat-topped on the 'Int' face, and as complementary depressions on the 'Ext' face (Fig. 58). They are somewhat larger in diameter than pinocytotic vesicles, and can be identified with certainty in situations where the narrow cytoplasm of the cell is revealed in cross-fracture between the two plasma membrane faces.

II NUCLEUS

The nucleus may present a variety of appearances in replicas. It may be directly cross-fractured so as to reveal an 'edge-on' view of the nuclear membrane, and the nucleoplasm, as in thin sections, or, an '*en face*' view of nuclear membrane with pores may be revealed (Fig. 59). Since

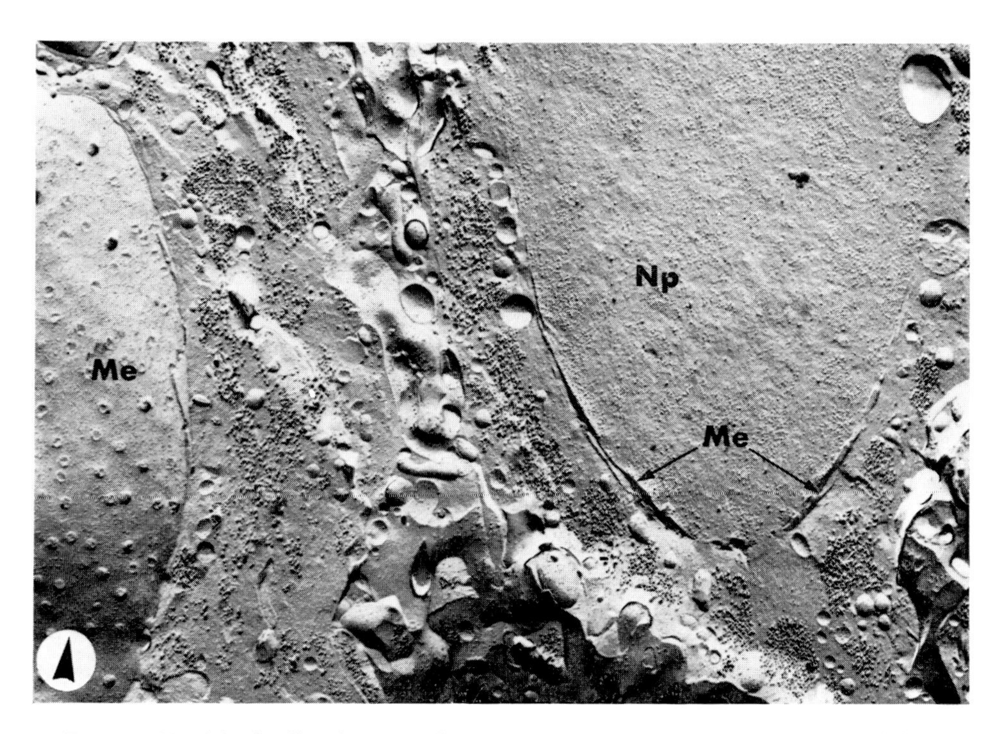

FIG. 59. Nuclei of cells of mature human epidermis. The nucleus on the right has been cross-fractured so as to reveal the nucleoplasm (Np) and an 'edge-on' view of the nuclear membrane (Me). An '*en face*' view of the membrane (Me) of the nucleus on the left has been revealed, and nuclear pores are prominent on the fracture face. × 16,000.

the nuclear membrane consists of two leaflets, outer and inner, complementary fracture faces of each leaflet can be revealed, directed either towards the exterior or the interior of the nucleus, i.e. four fracture faces in all (Figs. 59–62). Nuclear pores are evident on all faces, appearing either as depressions or as elevations, the particular appearance depending, apparently, upon whether or not the pore plug remains

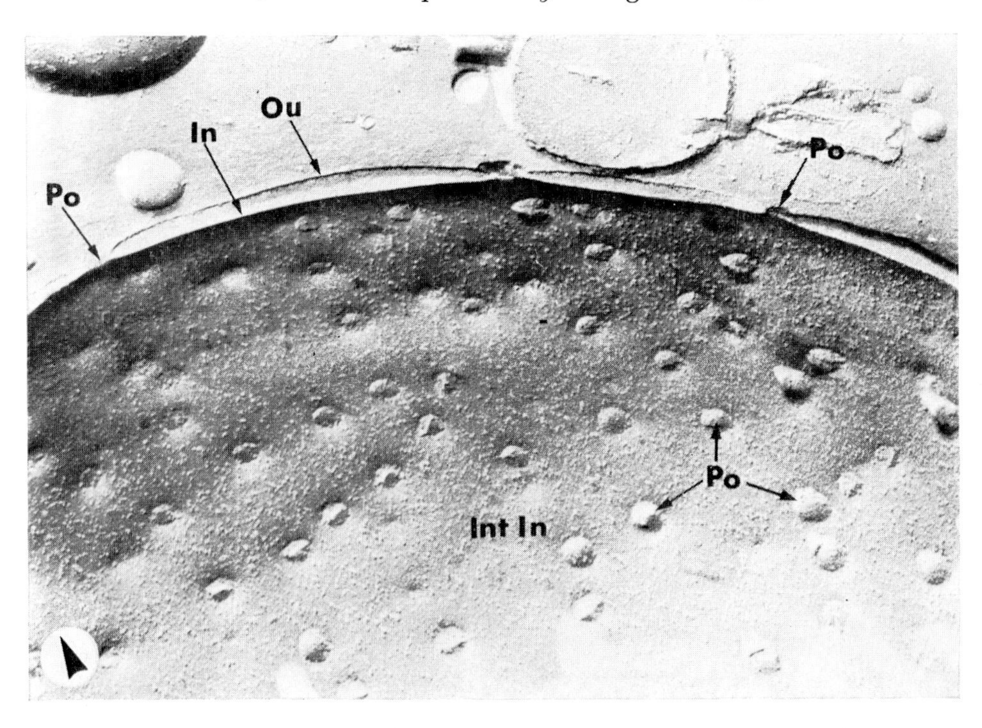

Fig. 60. This nucleus has been fractured so as to reveal edge-on views of the outer (Ou) and inner (In) nuclear membranes, and an extensive area of a fracture face of the inner membrane (Int In) directed towards the interior of the nucleus. Note nuclear pores (Po) on cross-fracture, and *en face*. × 37,500.

attached to an individual face on fracture (Monneron *et al.*, 1972). Maul (1971) reports on eight-fold symmetry (octagonality) in outline of individual pores revealed on the fracture face of the inner leaflet directed towards the interior of the nucleus.

It is evident from the figures that the freeze-fracture technique can permit realistic estimates of the concentration, arrangement, and distribution of nuclear pores. Until recently pores were thought to be randomly distributed over the nuclear surface but Maul *et al.* (1971) have shown that this is not the case, though they were not able to establish the exact distribution pattern. They also showed that the number of pores on the interphase nucleus can be altered by experimental procedures. This raises the possibility that numerical density, and other features of pores, may vary under physiological and pathological conditions. These are matters which seem worth investigating

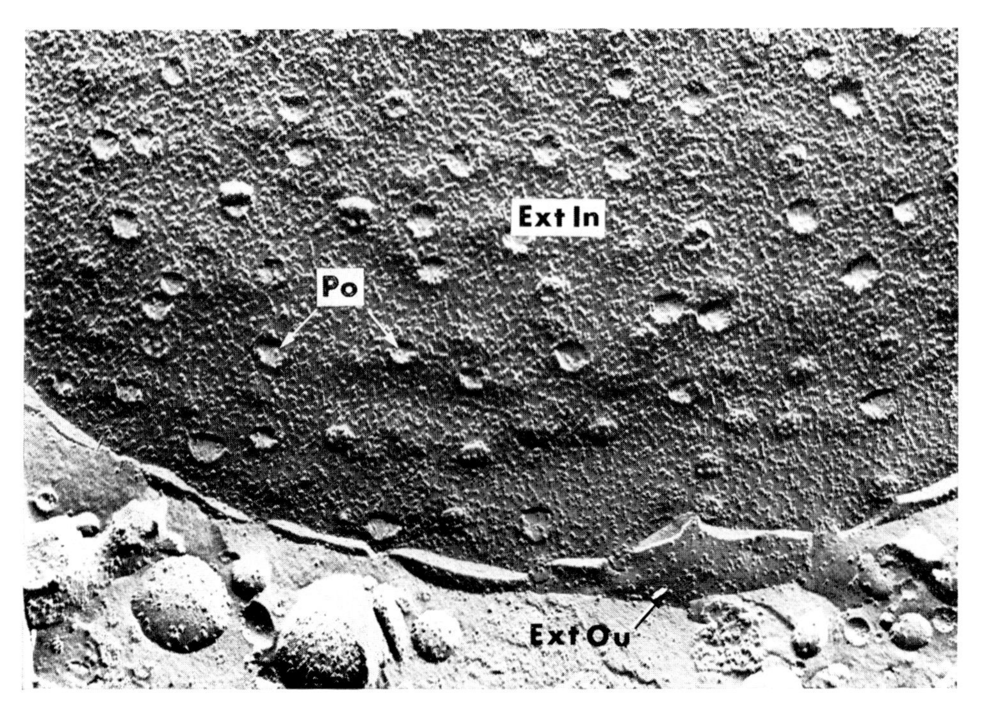

Fig. 61. Nucleus fractured so as to reveal Ext fracture faces of the outer (Ext Ou) and inner (Ext In) nuclear membranes. Po, nuclear pores. × 37,500.

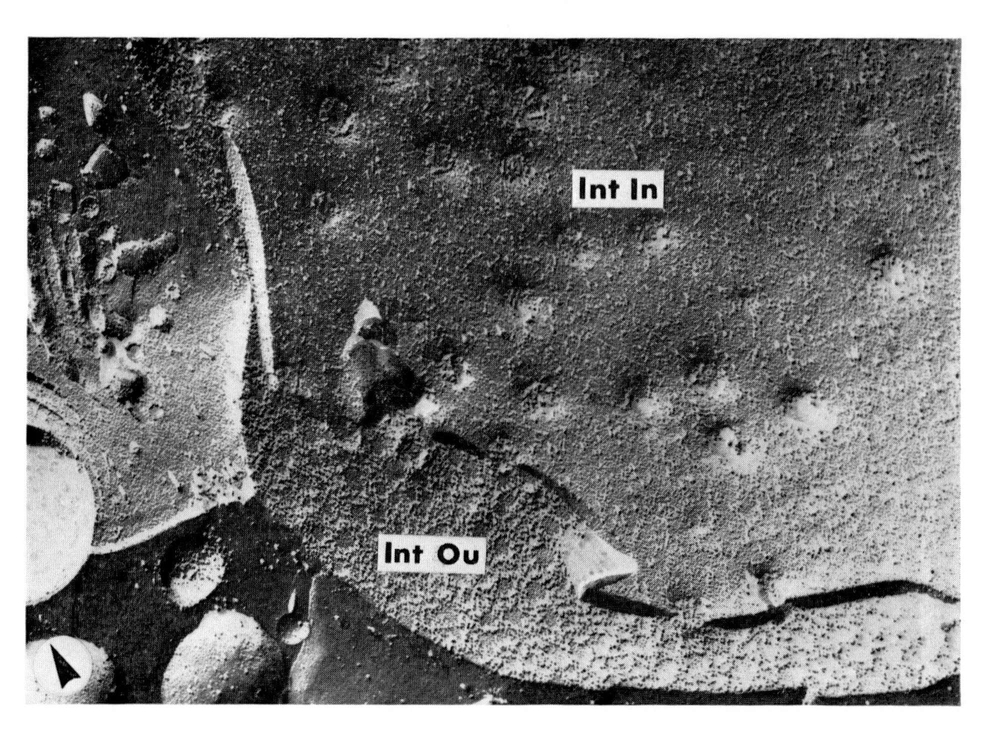

FIG. 62. Nucleus fractured so as to reveal Int fracture faces of the outer (Int Ou) and inner (Int In) nuclear membranes. × 37,500.

from the point of view of extending knowledge of normal nucleo-cytoplasmic information transfer, and its possible disturbance in specific disease conditions.

Wartiovaara and Branton (1970) have demonstrated ribosomes on the true outer surface of the nuclear membrane in freeze-fractured and sublimated material of sea-urchin embryos.

III CYTOPLASMIC ORGANELLES

A Endoplasmic Reticulum

When fractured transversely, the lamellae of cisternae of the rough endoplasmic reticulum present the highly characteristic appearance of stacks of undulating double ridges (Fig. 63). Where extensive areas of the lamellar membranes are revealed, two fracture faces separated by

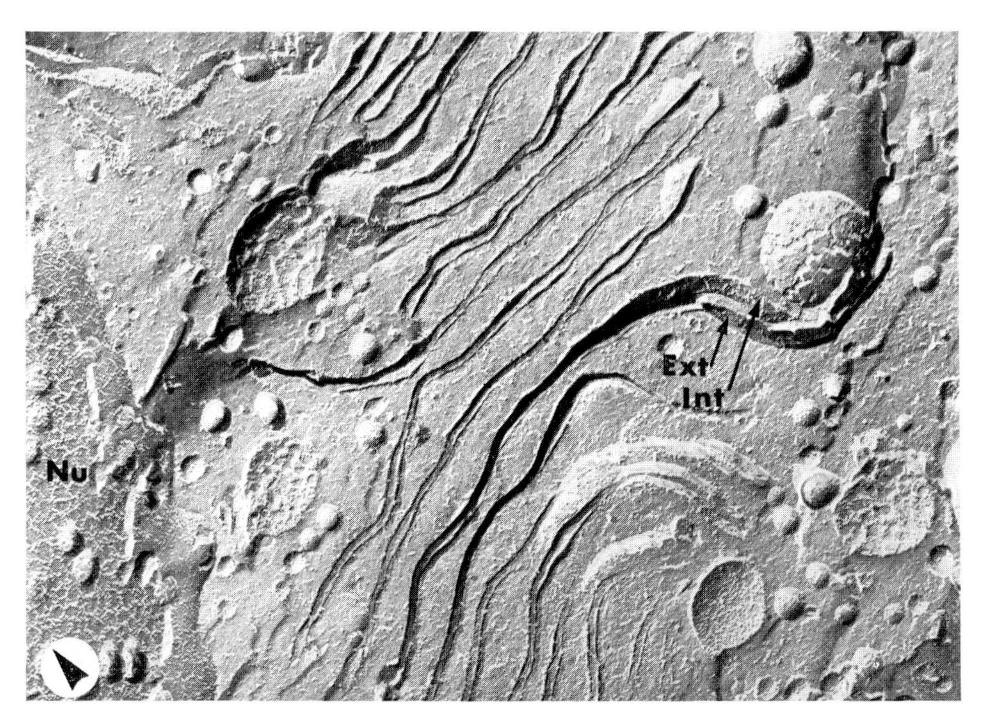

FIG. 63. Elongated stacks of undulating double ridges representing cross-fractured lamellar membranes of rough endoplasmic reticulum of mouse liver cell. Ext, Int, fracture faces of lamellar membrane (*see also* Fig. 64). Nu, nucleus. × 33,000.

a step which includes the cisternal lumen can be recognised (Figs. 63 and 64). One of these faces ('Ext') is directed towards the exterior, i.e. towards the cytoplasm, and the other ('Int') towards the interior, i.e. towards the cisternal lumen. These are evidently complementary fracture faces of the lamellar membrane, and, as may be seen from the figure, there are more particles on the 'Int' face than on the 'Ext' face.

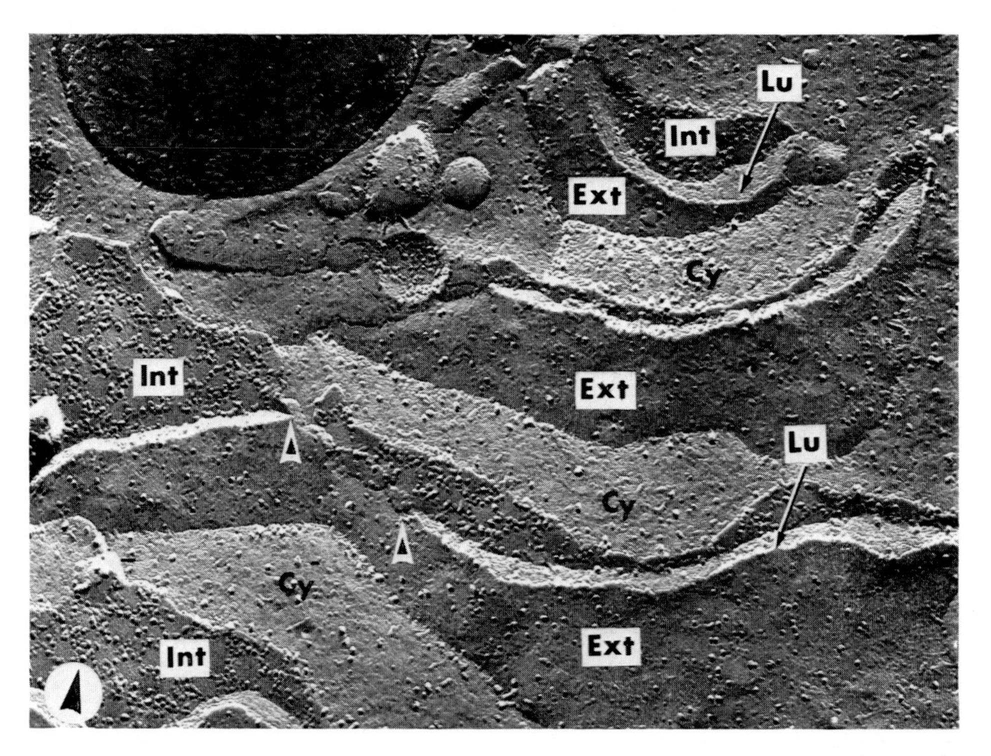

FIG. 64. Fracture faces of lamellar membranes of cisternae of rough endoplasmic reticulum of mouse liver cell. Cy, cytoplasm between cisternae; Ext, fracture face of cisternal membrane directed towards the exterior, i.e. towards the cytoplasm; Int, fracture face of cisternal membrane directed towards its interior, i.e. the lumen, (Lu). Note apparent obliteration of lumen, of largest cistern between large arrows due to presence of fenestrae, or pores. × 75,500.

Orci *et al.* (1972) have demonstrated fenestrae similar to nuclear pores on lamellar fracture faces of rough endoplasmic reticulum of exocrine pancreatic cells. They reported considerable variations in numbers of fenestrae from one cisterna to the other and mentioned the possibility that they may be highly labile structures, the prominence or otherwise of which may vary from species to species. Two such fenestrae are evident in Fig. 64.

Concentrations of small vesicles, and elongated elements present in cells which are known to contain a well developed smooth endoplasmic reticulum (Fig. 65) probably represent this system of membranes.

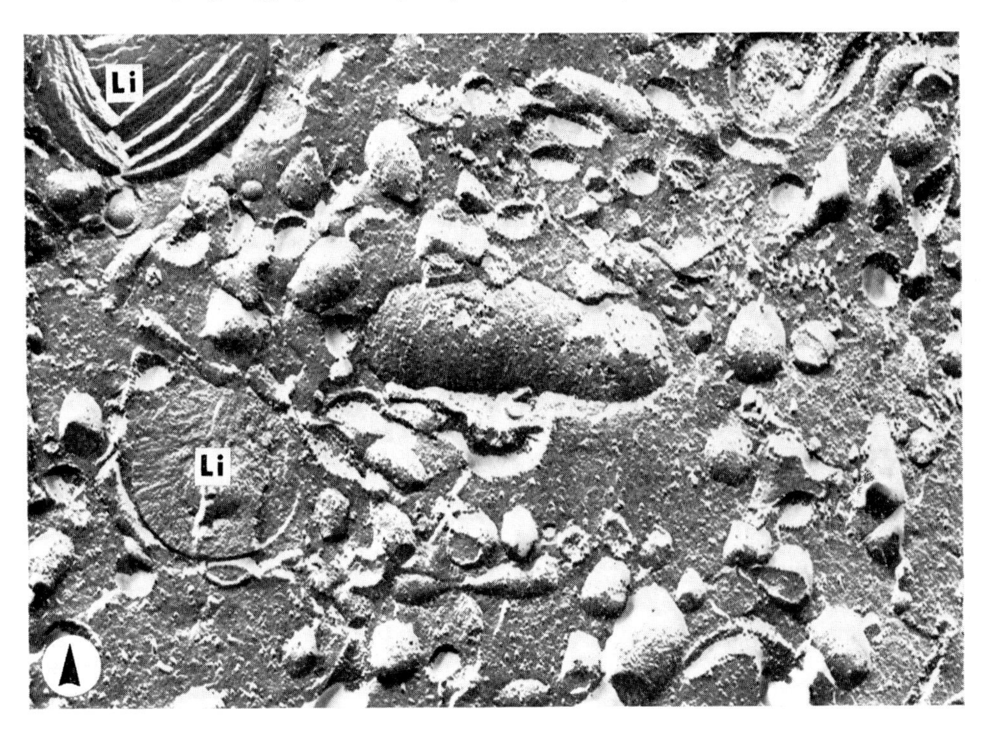

FIG. 65. Cytoplasm of cell of human sebaceous gland. These cells are known to contain extensive vesicles or tubules of the smooth endoplasmic reticulum, and the small rounded and elongated structures seen here may represent these. Li, lipid droplets. × 27,500.

B Mitochondria

With mitochondria of intact animal tissues, apart from cross-fractured images revealing cristae (Fig. 66), two types of membrane fracture images are most commonly seen. One of these 'Ext' (Fig. 67) which is directed towards the exterior of the organelle, has the appearance of raised plaques carrying the occasional particle separated by areas carrying many particles. The other type of image 'Int' (Fig. 67)

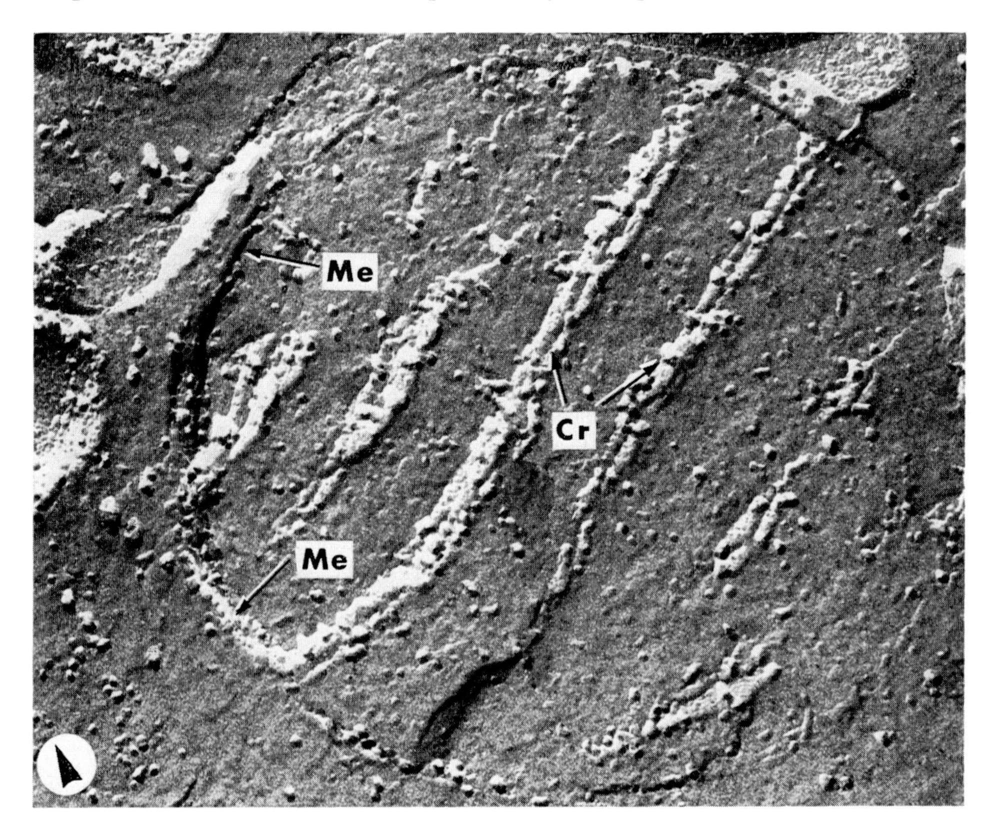

Fig. 66. Cross-fractured mitochondrion from mouse liver cell. Cr, internal cristae, evidently formed of double membranes; Me, limiting membrane of mitochondrion. × 165,500.

which is directed towards the interior of the organelle, has the appearance of smooth areas outlined by plaques and particles. There is some difficulty involved in interpreting these images. The two would appear to be complementary, and each probably represents a combination of fracture faces of the outer and inner mitochondrial membranes. Thus, the plaques on the first type 'Ext' (Fig. 67) could represent a fragmented fracture face of the outer membrane directed towards the exterior of the organelle, and the particle-covered areas a similarly directed face of the inner membrane. A similar explanation could be advanced to account for the appearance of the second type of fracture image 'Int' (Fig. 67), i.e., that it represents internally directed faces of the outer and

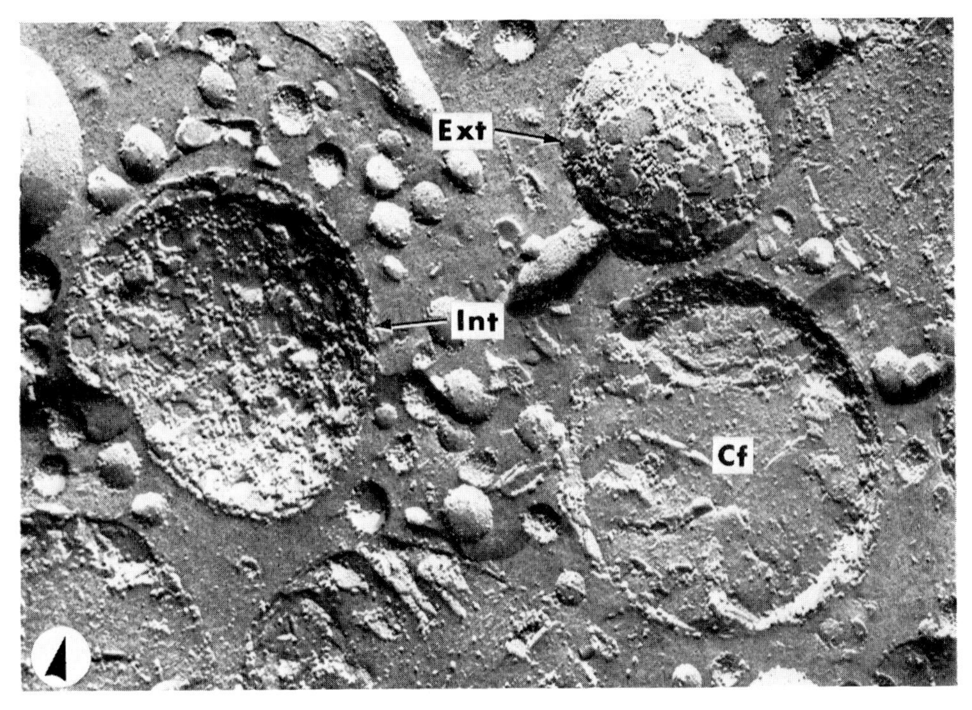

FIG. 67. Mitochondria from retinal pigment epithelial cell of guinea-pig. This shows two types of mitochondrial fracture faces commonly encountered with non-chemically fixed intact tissue, i.e. one (Ext) directed towards the exterior of the organelle, and another (Int) directed towards its interior. As suggested in the text, it is probable that both of these faces contain elements of outer and inner mitochondrial membranes. Cf, cross-fractured mitochondrion. × 66,000.

inner membranes. If this be the case, from the partial features revealed, one can build up a picture of the overall intact appearance of the respective fracture faces, e.g. of the externally directed face of the outer membrane as a relatively smooth one with few particles, and of the similarly directed face of the inner membrane as one carrying many particles (Fig. 68). This picture agrees quite well with descriptions of fracture faces of isolated intact mitochondria (Wrigglesworth *et al.*, 1970) and of separated outer and inner membranes (Melnick and Packer, 1971). It would appear that the inner and outer membranes of mitochondria of intact tissue are more adherent for some reason than those of isolated mitochondria so that on fracturing, the four expected faces are only partially revealed. Occasionally, isolated fracture faces

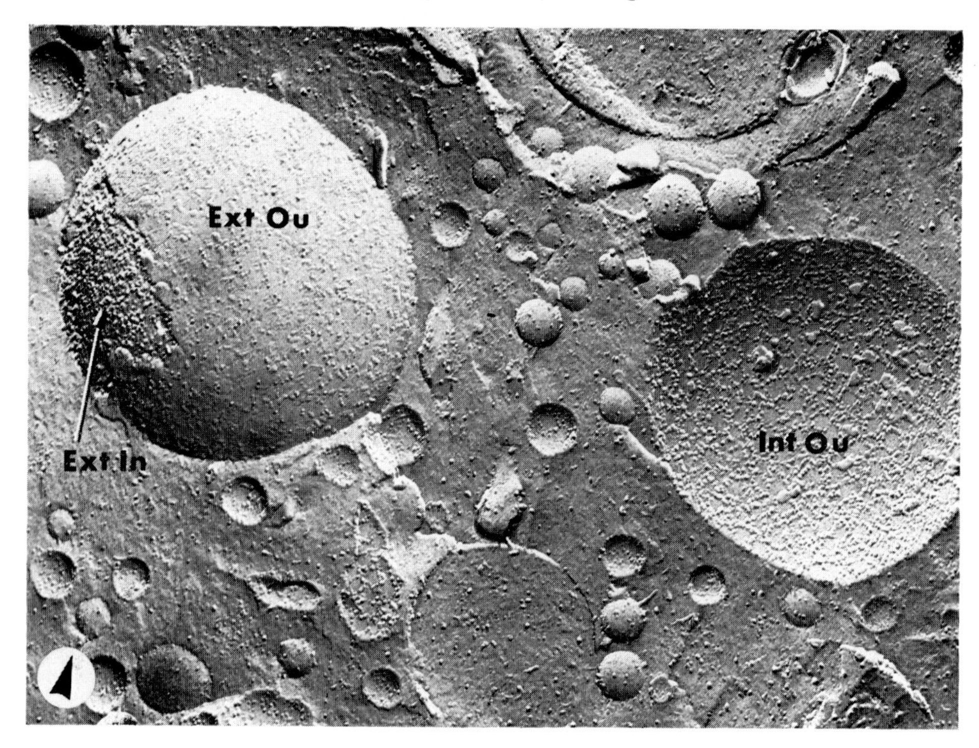

FIG. 68. Mitochondria from mouse liver cell. The organelle on the left is interpreted as revealing an almost complete Ext face of the outer membrane (Ext Ou) overlying a corresponding face of the inner membrane (Ext In) which carries many particles. The face on the right (Int Ou) is interpreted as an Int face of the outer mitochondrial membrane, though one cannot be certain of this identification. × 63,000.

which could represent one or other of the four in its entirety are seen (Fig. 68) but it is impossible to be certain whether these represent mitochondria or other vesicular elements of the cytoplasm.

C Golgi Apparatus

The Golgi apparatus appears in replicas as a stacked arrangement of curved membranes and vesicles in the para-nuclear cytoplasm (Fig. 69). It can quite easily be confused with elements of the rough endoplasmic reticulum, and indeed position is the best guide to its location. Being restricted to a particular region of the cell, one would not expect to see it in the majority of replicas.

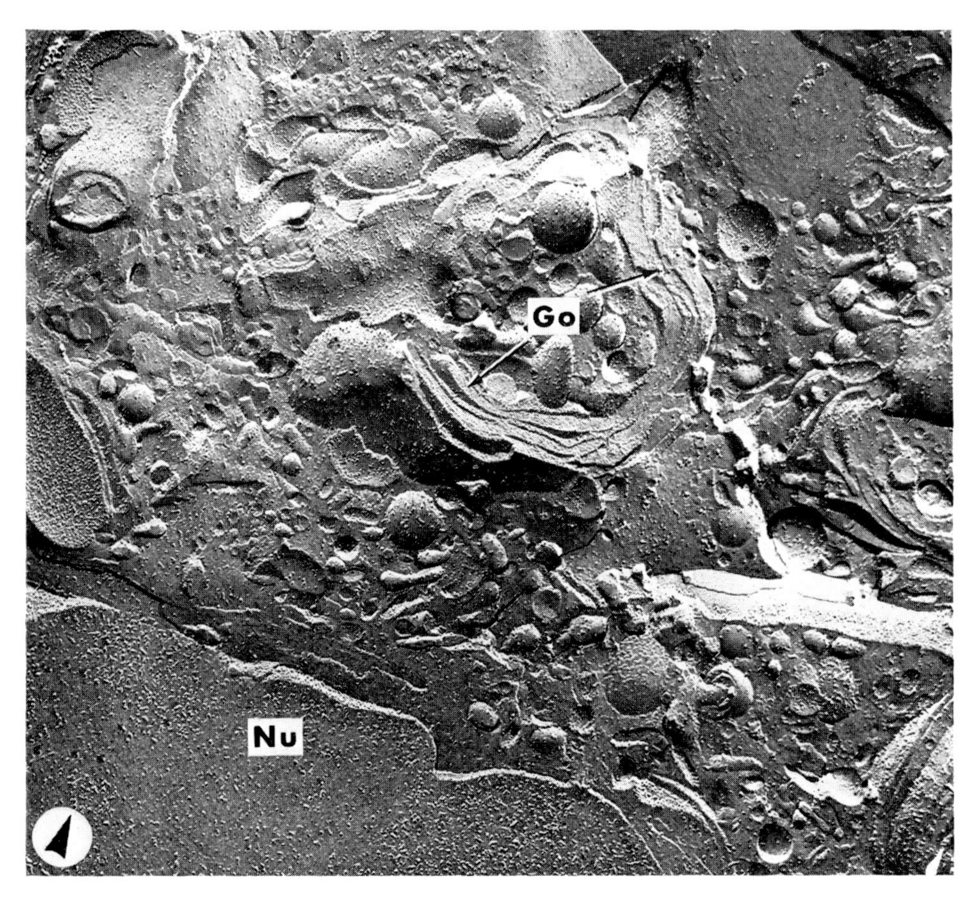

FIG. 69. Golgi apparatus (Go) in juxtra-nuclear (Nu) region of cytoplasm of epithelial cell of mouse jejeunum. × 34,000.

D Other General Cytoplasmic Elements

1 Lipid. Lipid or fat droplets appear in replicas as multi-lamellar structures of varying size and the overall appearance of individual droplets varies, depending upon the way the lamellae are fractured (Figs 70 and 71). Freeze-fracture confirms the fact that lipid droplets lie free in the cytoplasm, and are not surrounded by a limiting membrane (Hasegawa and Uyeda, 1974).

FIG. 70. Lipid droplets (Li) in cytoplasm of mouse liver cell. Note extensive lamination of droplet on the left. The absence of particles from the face of the droplet on the right leads to the conclusion that it lies free in the cytoplasm and is not membrane-bound. × 110,000.

2 Cytofilaments. Cytofilaments are a regular feature of many cell types but perhaps most characteristically of the mammalian epidermal keratinocyte. In this situation, as indicated by the length of shadows cast by them, fractured filaments project somewhat beyond the

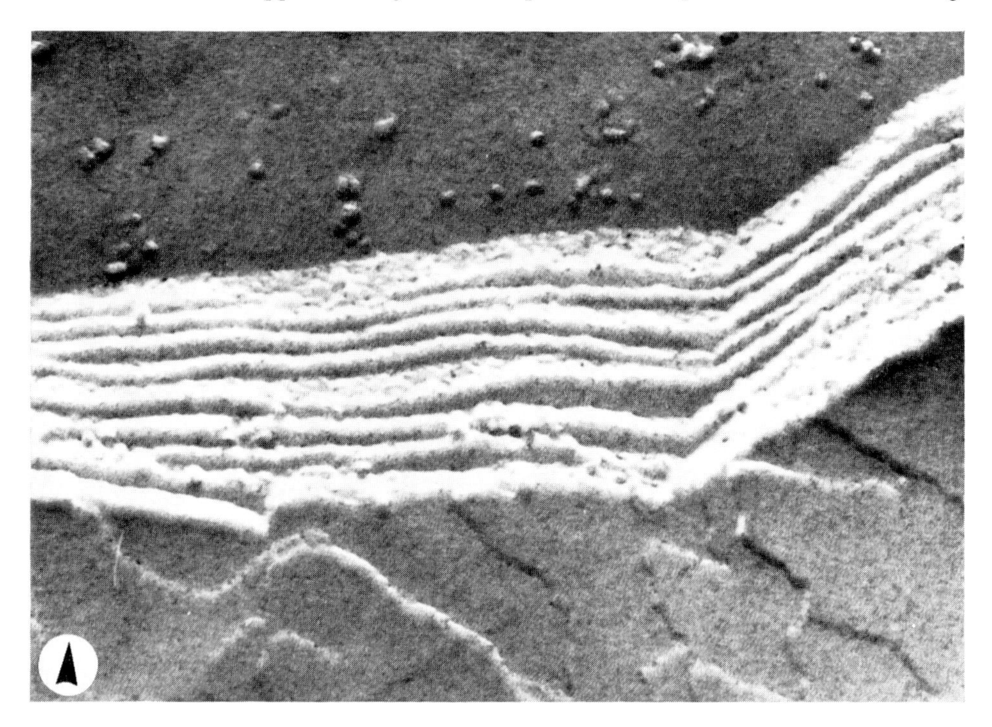

FIG. 71. *En face* and edge-on views of fractured lamellae of lipid droplet from retinal pigment epithelium of guinea-pig. × 254,000.

general plane of fractured cytoplasm (Fig. 72). This suggests that filaments are pulled slightly out of the matrix, or stretched, before they actually fracture. In general, the concentration of epidermal filaments seen in replicas is significantly less than in thin sections. This could be due to the fact that the other ends of stretched or pulled-out filaments would lie below the level of the general fracture plane and therefore would not be visualised.

3 Glycogen. It is practically impossible to be certain of the identification of individual glycogen particles distributed at random throughout the cytoplasm. However, in certain situations (Fig. 73) where heavy concentrations of glycogen are known to be present to the virtual exclusion of other cytoplasmic organelles, particles are seen which presumably represent this polysaccharide store.

4. Lysosomes. Apart from vesicular or spherical elements which can confidently be identified as belonging to one or other of the above

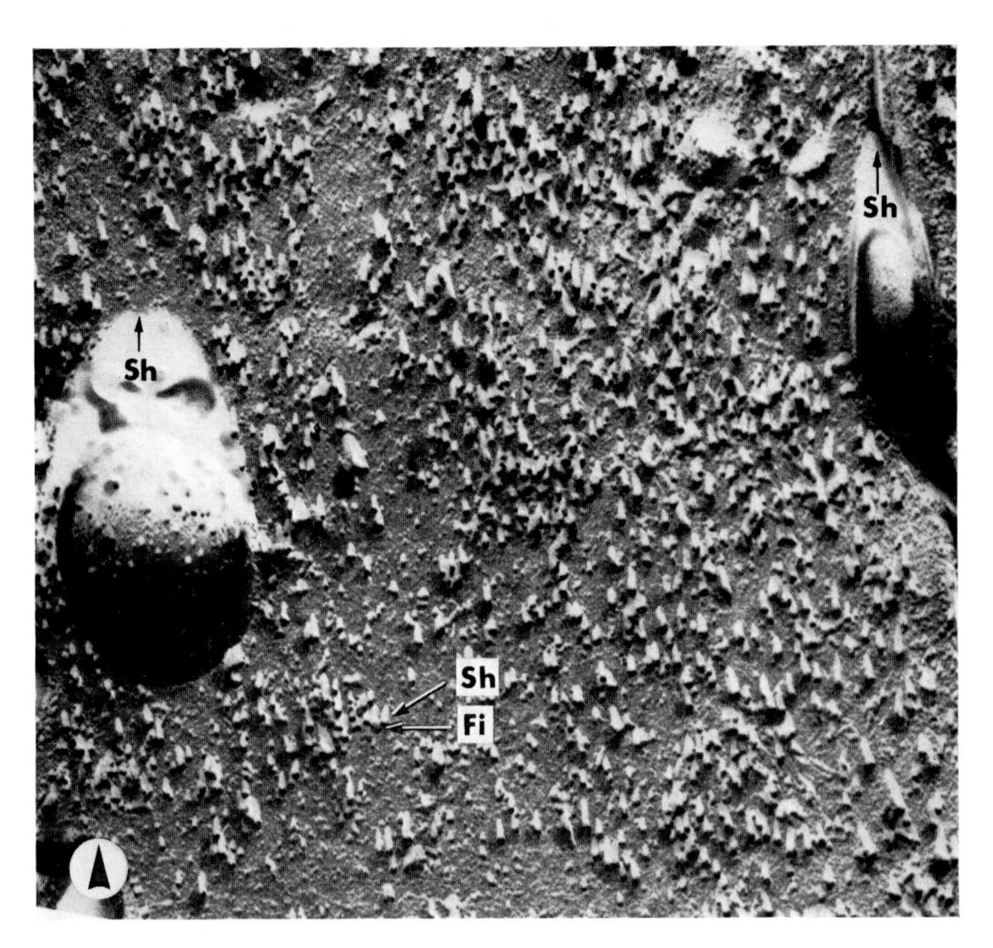

Fig. 72. Filaments (Fi) and associated shadows (Sh) in cytoplasm of granulosa-layer cell of mature human epidermis. Note also shadows cast by 'convex' cytoplasmic organelles. × 67,000.

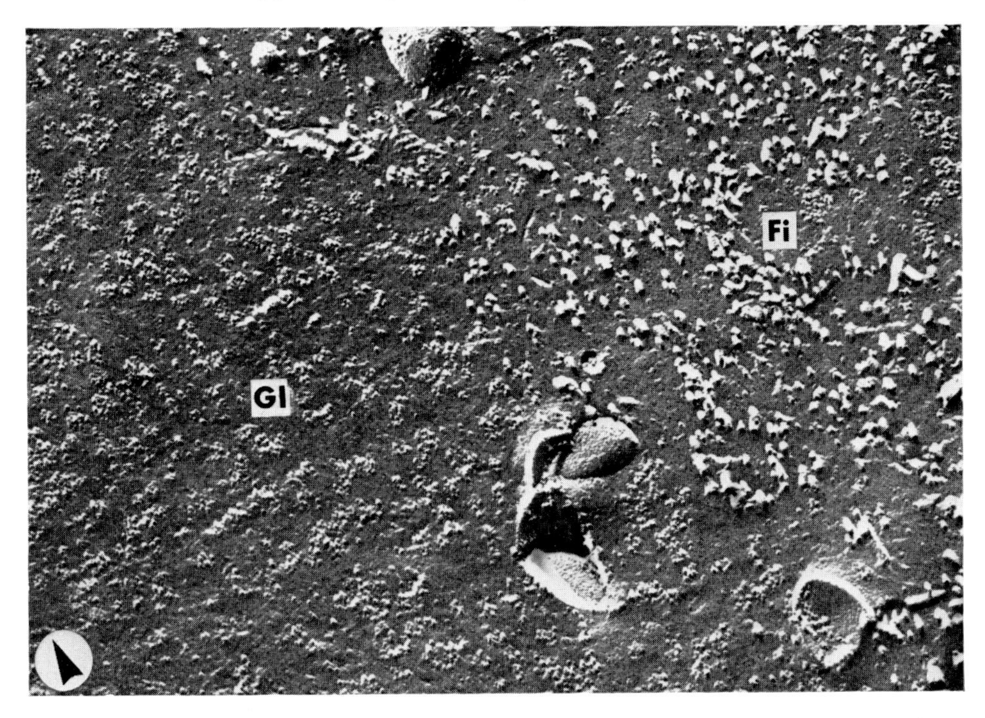

FIG. 73. Replica of cytoplasm of periderm cell of human foetal epidermis. These cells are known to contain large concentrations of glycogen particles adjacent to areas of filaments. Filaments (Fi) can be identified readily, and it is assumed that the area marked Gl represents glycogen. × 56,500.

described structures, every replica exhibits some which it is impossible to label with certainty. Some observers refer to these as lysosomes, mainly on grounds of elimination, but we feel that the freeze-fracture appearances of this variety of organelle remain to be clearly established, with the exception, possibly, of the multi-vesicular body (Fig. 97).

REFERENCES

Albertini, D. F. and Anderson, F. (1974). *J. Cell. Biol.* **63**, 234–250.
Breathnach, A. S. (1975). *J. Invest. Dermatol* (in press).
Breathnach, A. S., Stolinski, C. and Gross, A. (1972). *Micron* **3**, 287–304.
Breathnach, A. S., Goodman, T., Stolinski, C. and Gross, M. (1973). *J. Anat.* **114**, 65–81.
Chalcroft, J. P. and Bullivant, S. (1970). *J. Cell. Biol.* **47**, 49–60.

Farquhar, M. G. and Palade, G. E. (1963). *J. Cell Biol.* **17**, 375–412.
Flower, N. E. (1971). *J. Ultrastruct. Res.* **37**, 259–268.
Friend, D. S., and Gilula, N. B. (1972). *J. Cell Biol.* **53**, 758–776.
Hasagawa, T., and Uyeda, K. (1974). *Arch. Derm. Forsch.* **248**, 347–354.
McNutt, S. N. and Weinstein, R. S. (1970). *J. Cell Biol.* **47**, 666–688.
McNutt, S. N. and Weinstein, R. S. (1973). *In* 'Progress in Biophysics and Molecular Biology', **26** (J. A. V. Butler and D. Noble, eds) 47–101. Pergamon, Oxford.
Maul, G. M. (1971). *J. Cell Biol.* **51**, 588–593.
Maul, G. M., Price, J. W. and Liebermann, M. W. (1971). *J. Cell. Biol.* **51**, 405–418.
Melnick, R. L. and Packer, L. (1971). *Biochem. Biophys. Acta* **253**, 503–508.
Mercer, E. H. (1964). *In* 'The Epidermis' (W. Montagna and W. C. Lobitz, eds) 161–178. Academic Press, New York.
Monneron, A., Blobel, G. and Palade, G. E. (1972). *J. Cell Biol.* **55**, 104–125.
Orci, L., Perrelet, A. and Libe, A. A. (1972). *J. Cell. Biol.* **55**, 245–259.
Simionescu, M., Simionescu, N. and Palade, G. E. (1974). *J. Cell Biol.* **60**, 128–152.
Spycher, M. A. (1970). *Z. Zellforsch.* **111**, 64–74.
Staehelin, L. A. (1973). *J. Cell Sci.* **13**, 763–786.
Staehelin, L. A., Mukherjee, T. M. and Williams, A. W. (1969). *Protoplasma* **67**, 165–184.
Wartiovaara, J. and Branton, D. (1970). *Exptl. Cell Res.* **61**, 403–406.
Wrigglesworth, J. M., Packer, L. and Branton, D. (1970). *Biochim. Biophys. Acta* **205**, 125–135.

6

Cells and Organised Tissues

In this chapter micrographs of a selection of individual cells and tissues are presented in order to demonstrate the general application of the technique, and where appropriate, its particular advantages in revealing structural features not readily demonstrable by other techniques. This section is not intended to represent a comprehensive atlas of any of the cells or tissues concerned. Indeed some of them are, as yet, virtually unexplored by the technique, and readers who may be disappointed in finding that their own favourite tissue is not featured at all, must accept our apologies. Deficiencies in this respect may be corrected in future editions of the book.

I UNICELLULAR ORGANISMS

A Bacteria (Figs. 74–79)

Giesbrecht (1966) and Remsen and Lundgren (1966) were the first to apply freeze-fracture to the examination of bacteria, and since then a variety of organisms have been examined (Remsen and Watson, 1973; Nanninga, 1973). Some of the earlier studies were of fundamental significance in illuminating the problems of the plane of fracture of biological membranes in general (Nanninga, 1971), and the technique has proved particularly valuable for the study of bacterial cell envelopes, cytomembranes, and storage products (Watson and Remsen, 1970; Kirk and Ginzberg, 1972; Lickfeld *et al.*, 1972). Each bacterium has its own particular characteristics, and here we can only illustrate some general features by presenting pictures of one example. We have chosen *Rhodopseudomonas spheroides*, NCIB, 8253, a Gram-negative photosynthetic organism. When examined by routine transmission electron microscopy, this bacterium exhibits a cell wall, a plasma membrane, and a cytoplasm which contains chromatophores, and occasionally, storage granules. It has no nucleus, and very rarely flagella.

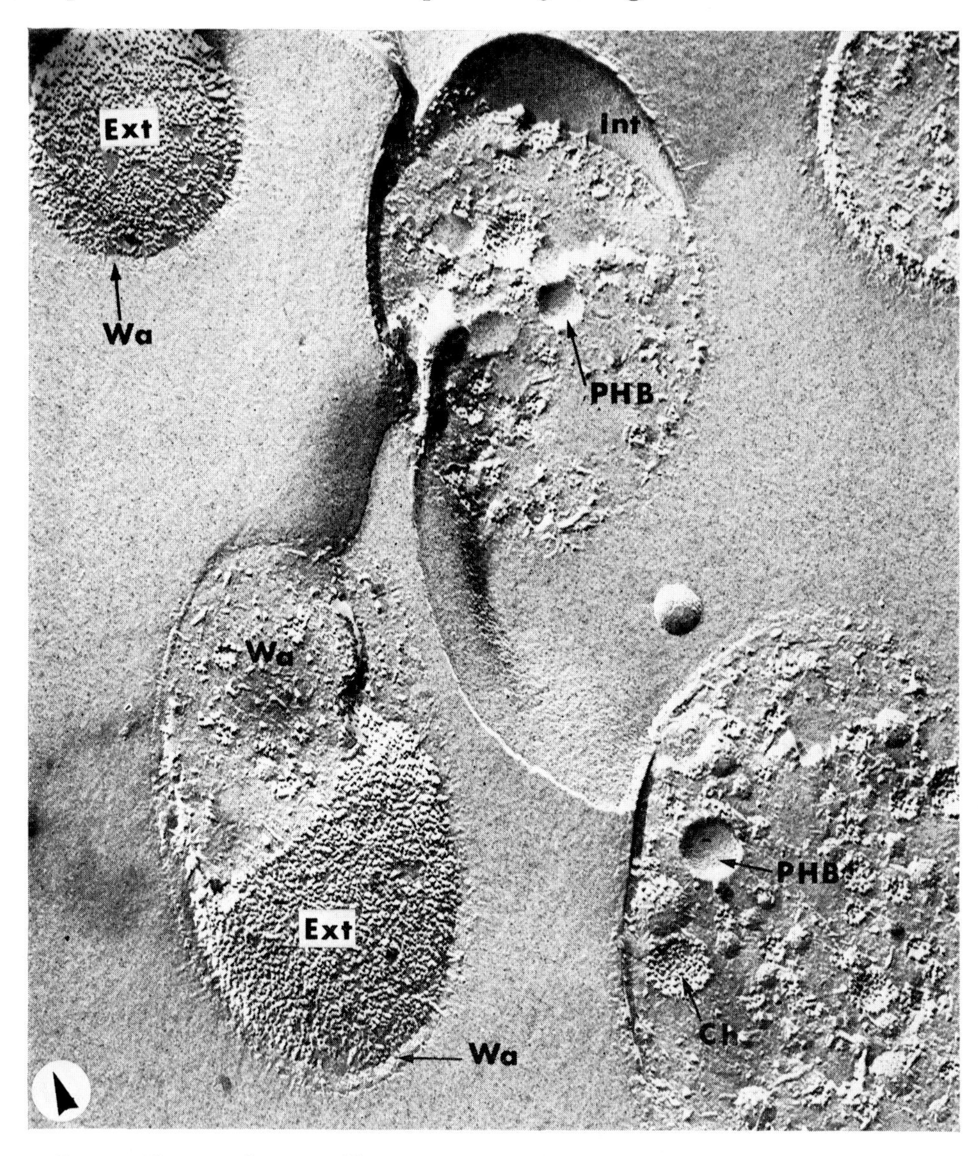

FIG. 74. Fracture images of bacterium, *Rhodopseudomonas spheroides*. Ch, fracture face of chromatophore membrane; Ext, fracture face of plasma membrane directed towards exterior of organism; Int, fracture face of plasma membrane directed towards interior of organism; PHB, depression left by PHB granule which has been fractured away; Wa, wall. × 66,000.

Most of these features are evident on one or other of the accompanying illustrations (Figs. 74–79). Hexagonal patterns are frequently observed on the cell wall of bacteria (Watson and Remsen, 1970), but not in this instance. Perhaps they might become evident on sublimation. 'Ext' and 'Int' type fracture faces of the plasma membrane are seen, the latter, carrying the greater number of particles (Figs. 74 and 75). A striking feature of these faces is the occurrence of a circumscribed area devoid of particles (Figs. 75 and 76). This could be artefactual, but it is of interest to note that a similar area has been observed on fracture faces of the plasma membrane of *Holobacterium* (Kirk and Ginzberg, 1972).

The cytoplasm exhibits fracture faces of chromatophores of varying diameters, depending upon how close or otherwise to the equator the

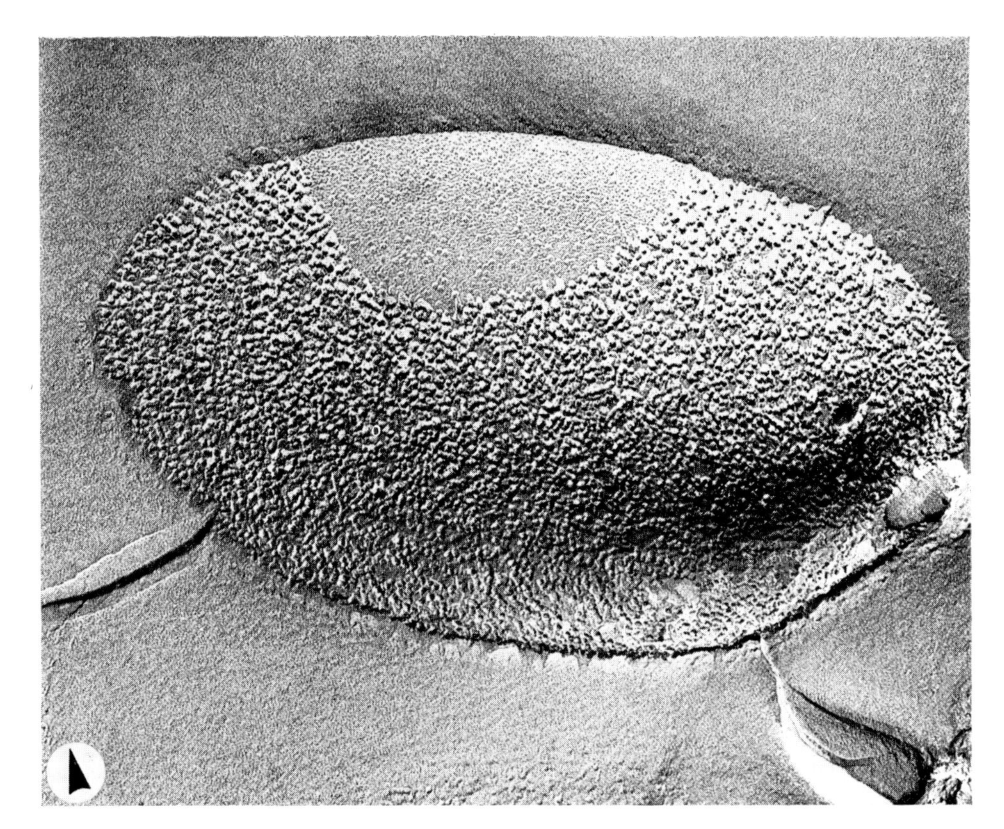

FIG. 75. *Rhodopseudomonas spheroides*, The Ext fracture face of the plasma membrane is revealed. Note localised area devoid of particles. × 99,000.

FIG. 76. *Rhodopseudomonas spheroides.* Int fracture face of plasma membrane complementary to the Ext face seen in Fig. 75. × 84,000.

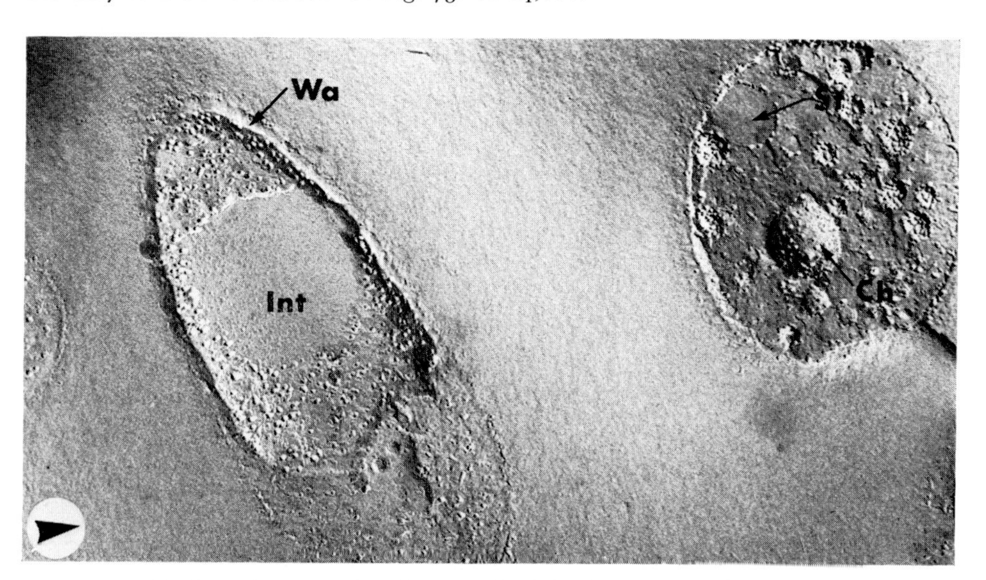

FIG. 77. *Rhodopseudomonas spheroides.* On the left, Int fracture face of plasma membrane and wall of one organism. On the right, a cross-fractured specimen with chromatophore membranes (Ch) and storage granule (St). × 66,000.

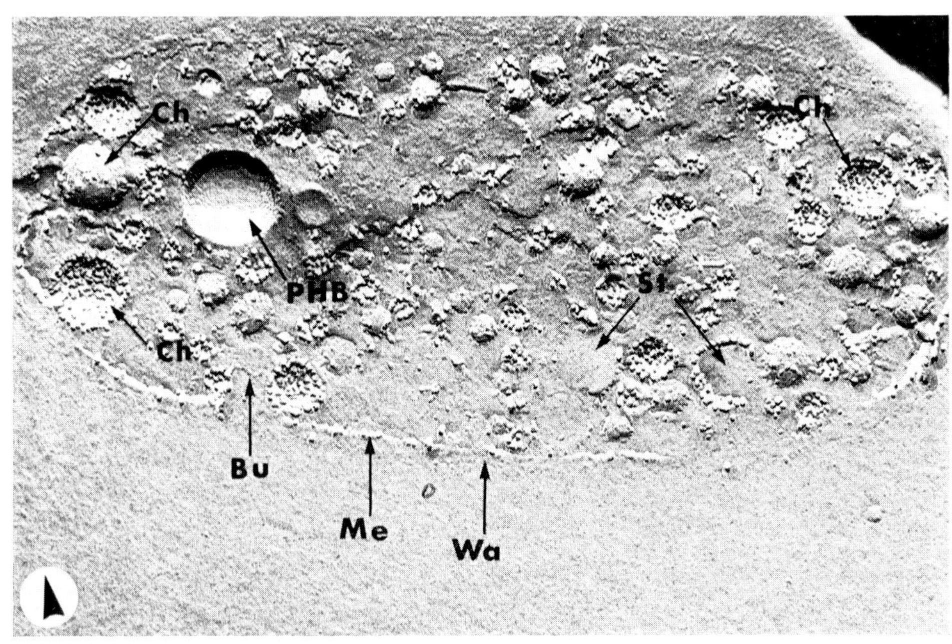

FIG. 78. Longitudinally fractured specimen of *Rhodopseudomonas spheroides*. Bu, inward bulging of plasma membrane (Me) which could represent forming chromatophore; Ch, fracture faces of chromatophore membrane; PHP, hollow left by PHB granule which was fractured away; St, storage granule, Wa, wall. × 70,000.

FIG. 79. *Rhodopseudomonas spheroides*. PHB, granule stretched and deformed during fracturing. The extent to which it projects beyond the general fracture plane can be judged by the length of the shadow. × 132,000.

fracture entered the organelle. Occasionally, what may be interpreted as chromatophore membranes in continuity with, or budding off from the plasma membrane, may be seen (Fig. 78). PHB granules are represented by apparently plastically deformed elevations or 'horns' and by depressions from which granules are removed during the process of fracturing (Fig. 79). Other storage granules usually appear as mem-

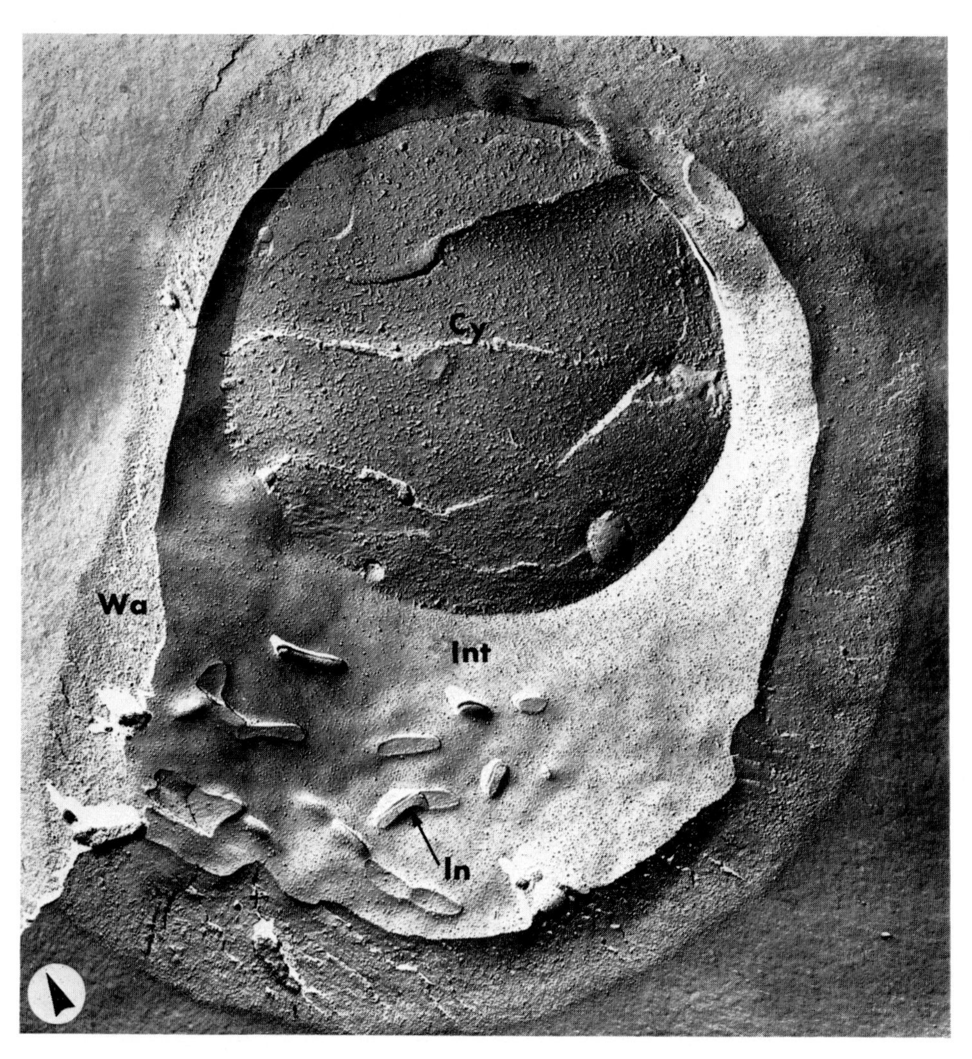

FIG. 80. Replica of yeast cell showing cross-fractured wall (Wa), cytoplasm (Cy) with one mitochondrion, and fracture face of plasma membrane directed towards the interior of the cell (Int). In, infolding of plasma membrane. × 30,500.

brane-limited bodies with smooth internal structure, and some are tentatively identified in the figures.

B Yeast (Figs. 80–82).

A classic freeze-fracture study of yeast cells was published at a comparatively early stage by Moor and Mühlethaler (1963), and little of essential significance has been added to their observations since then. Cell wall, plasma membrane, and cytoplasm are well shown (Figs. 80–82). A characteristic feature of the 'Ext' face of the plasma membrane is the occurrence of hexagonally arranged particles from which fibrils extend into the cell wall (Fig. 81). Infoldings of the plasma membrane are clearly shown on both fracture faces. The cytoplasm

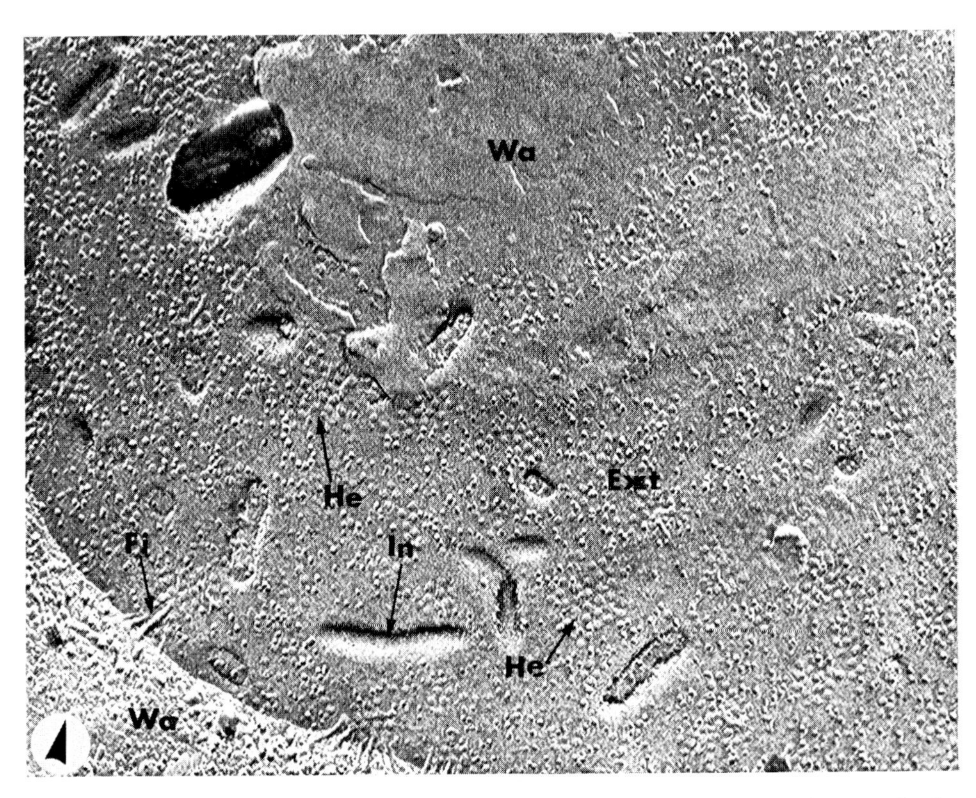

Fig. 81. Fracture face of plasma membrane of yeast cell directed towards the exterior (Ext). Note hexagonal arrangement of particles (He) and fibrils (Fi) extending from particles to cell wall (Wa). × 104,500.

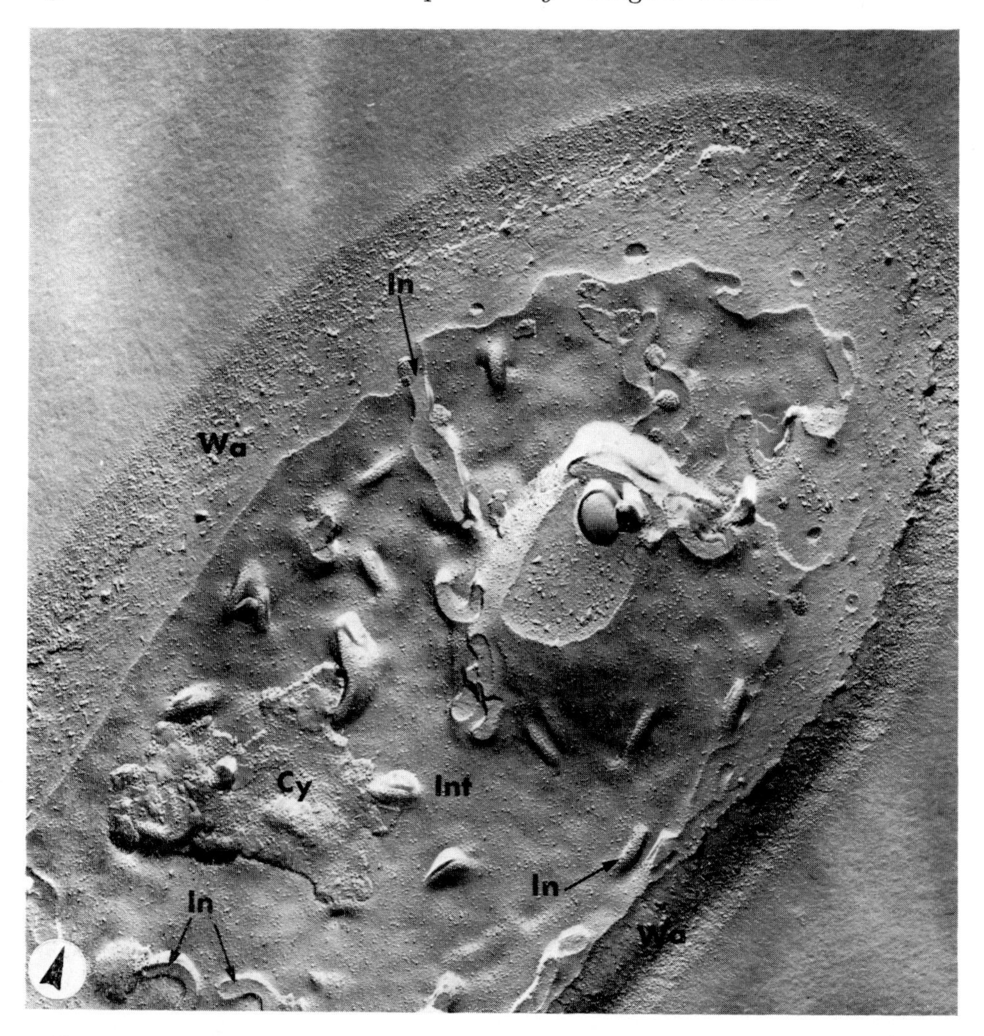

Fig. 82. Int fracture face of plasma membrane of yeast cell. Cy, cytoplasm; In, infoldings of plasma membrane; Wa, cell wall. × 38,000.

contains the common organelles – mitochondria, endoplasmic reticulum, and in particular, large droplets of lipid. These types of organelles have been sufficiently illustrated elsewhere so as not to require representation here.

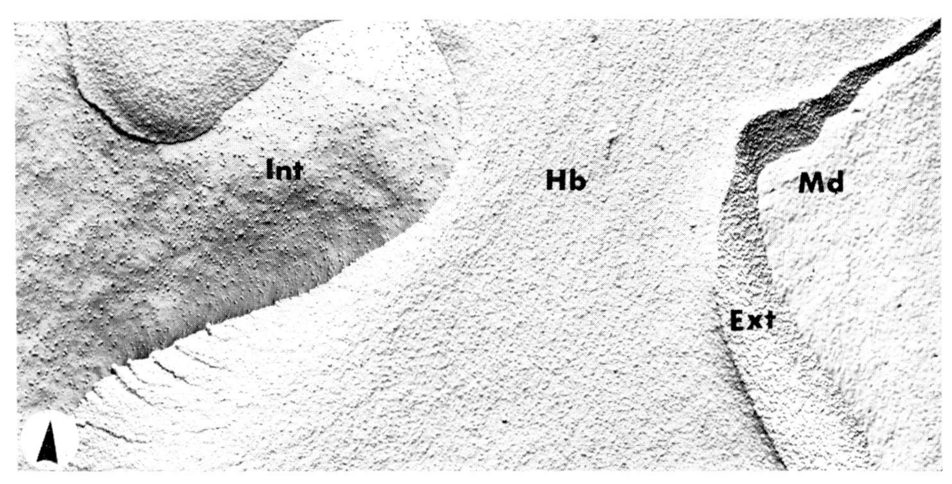

FIG. 83. Replica of portion of freeze-fractured human erythrocyte. Ext, fracture face of plasma membrane directed towards the exterior, in this case, the suspending buffered glycerol-plasma medium, Md; Int, fracture face of plasma membrane directed towards the interior; Hb, Haemoglobin. × 30,500. For other pictures of erythrocytes, see Figs. 8, 38 and 42.

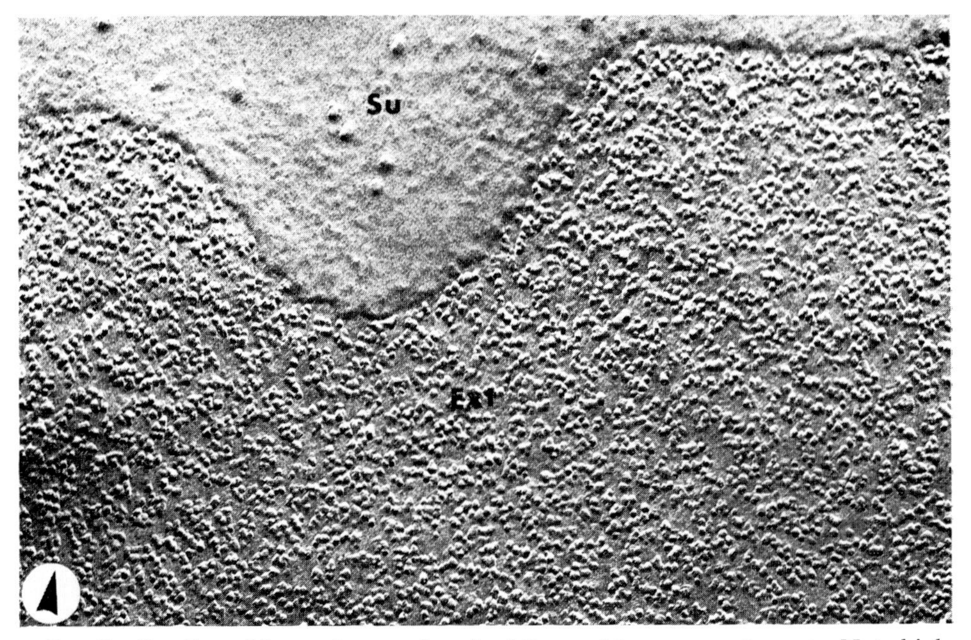

FIG. 84. Replica of freeze-fractured and sublimated human erythrocyte. Note high concentration of particles on Ext face of plasma membrane. Su, true outer surface of plasma membrane revealed by sublimation. × 105,000.

II BLOOD CELLS AND VASCULAR ENDOTHELIUM (FIGS. 83–90)

The erythrocyte is one of the most convenient and readily available of specimens for biological investigation. Weinstein and Bullivant (1967) and Huhn and Grassmann (1969) were among the first to apply freeze-fracture techniques to the examination of the erythrocyte, demonstrating the typical fracture faces of the plasma membrane and the membrane-associated particles (Figs. 83–85). Subsequently Pinto da Silva and Branton (1970), Tillack and Marchesi (1970), and Branton (1971) performed elegant sublimation and labelling experiments on erythrocyte ghosts suspended in distilled water which greatly advanced understanding of the plane of fracture of membranes in general, and the nature of membrane-associated particles (see Chapter 4). This fundamental work on the erythrocyte provides a sound basis for the application of the technique towards mapping specific sites on the erythrocyte membrane, a matter of interest for the immunologist and haematologist.

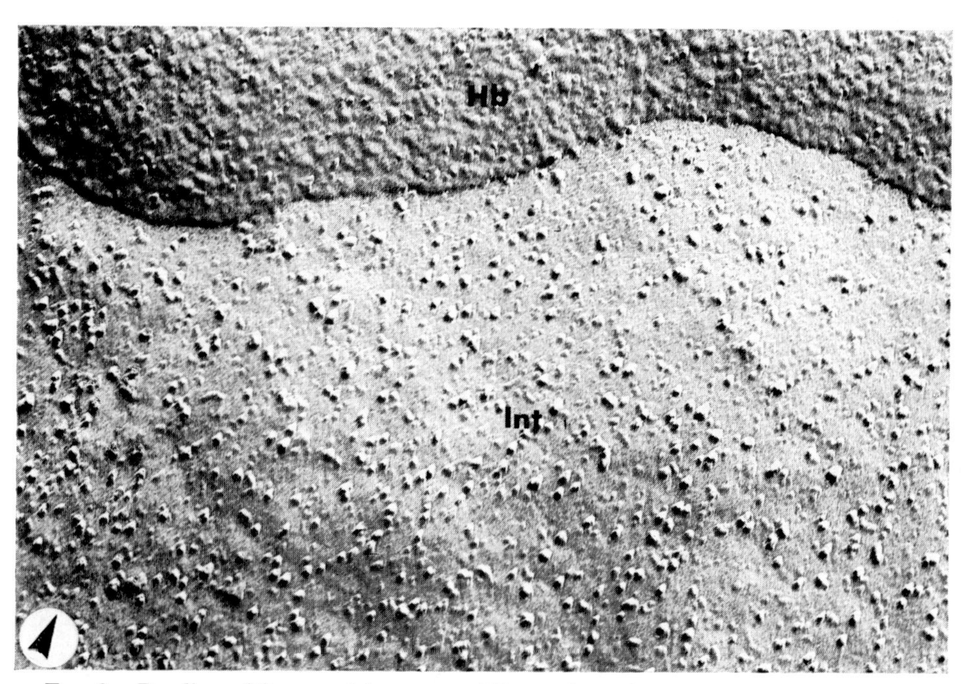

FIG. 85. Replica of fractured but not sublimated erythrocyte. Note fewer particles on Int fracture face of plasma membrane as compared with Ext face in Fig. 84. Hb, haemoglobin. × 102,000.

Granulocytes present a highly characteristic appearance in replicas (Fig. 86). Being lobed, the nucleus frequently appears fractured in

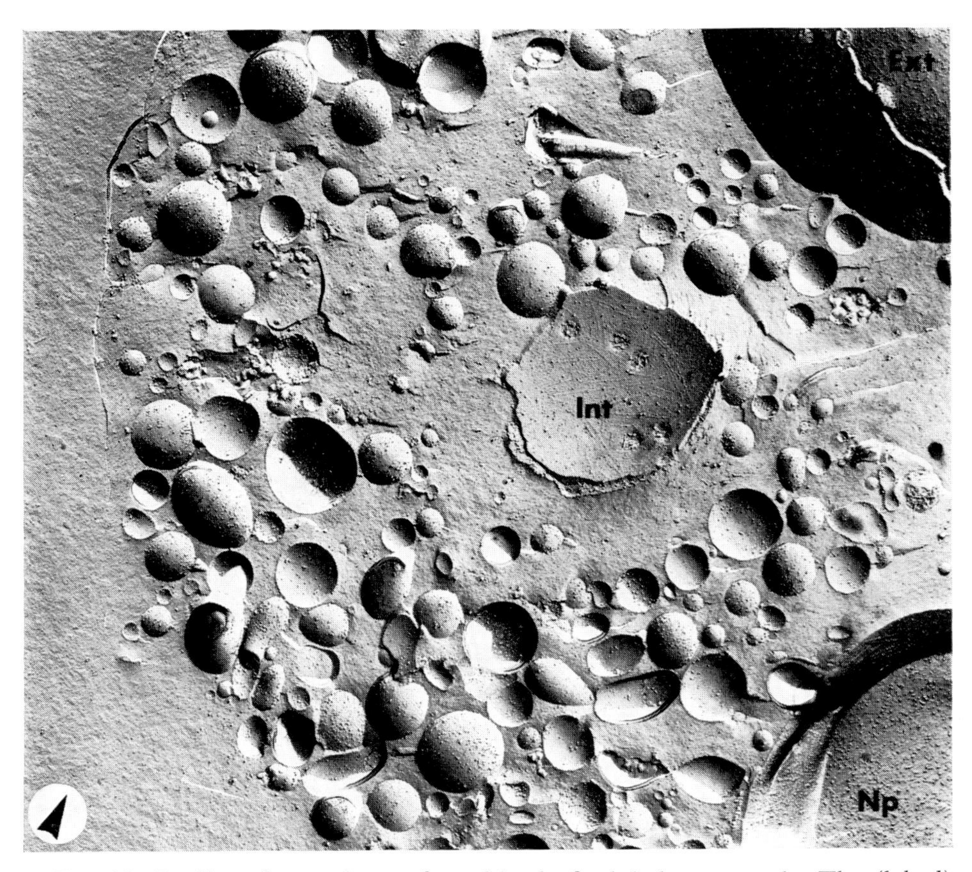

FIG. 86. Replica of granulocyte from blood of adult human male. The (lobed) nucleus appears fractured in three locations exhibiting Ext and Int fracture faces of both leaflets of the membrane, and the nucleoplasm (NP). × 30.000.

several places, and the cytoplasm contains numerous granules, the majority of which fracture so as to reveal membrane faces rather than internal contents. Pseudopodia or processes projecting from the plasma membrane provide a useful guide for identifying lymphocytes (Fig. 87). Platelets have been described by Hoak (1972).

Vascular endothelial cells are receiving increasing attention because the freeze-fracture technique has unique advantages in demonstrating specialised contacts, fenestrae or pores (Figs. 88–90), and variations in the distribution of other subcellular structures (Smith *et al.*, 1973; Simionescu *et al.*, 1974). The latter authors provide data concerning the dimensions and frequency of fenestrae and micropinocytotic vesicles which can be related to variations in capillary permeability.

FIG. 87. Replica showing Int fracture face of plasma membrane of lymphocyte from blood of adult human male. The processes or pseudopodia (Pr) are characteristic of the lymphocyte. × 20,000.

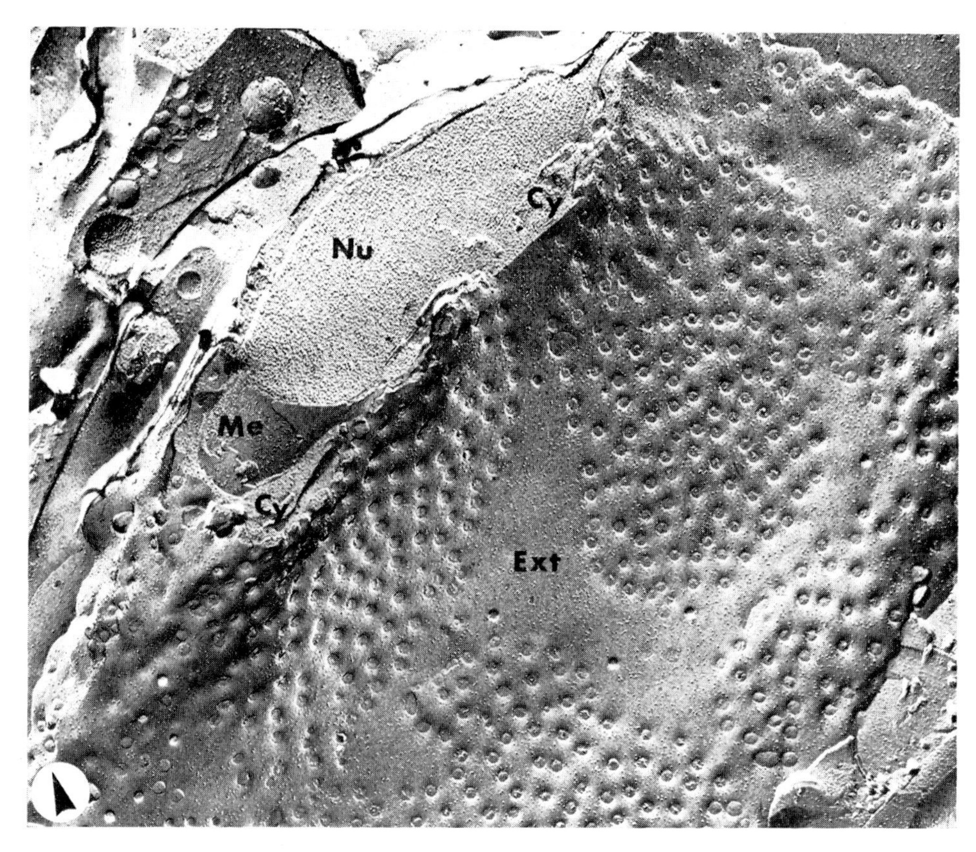

Fig. 88. Replica revealing in the main, Ext fracture face of plasma membrane of capillary endothelial cell of mouse jejunum. Note smooth areas alternating with areas of fenestrae on the fracture face. Cy, cytoplasm; Me, fracture face of nuclear membrane; Nu, nucleoplasm. × 27,000.

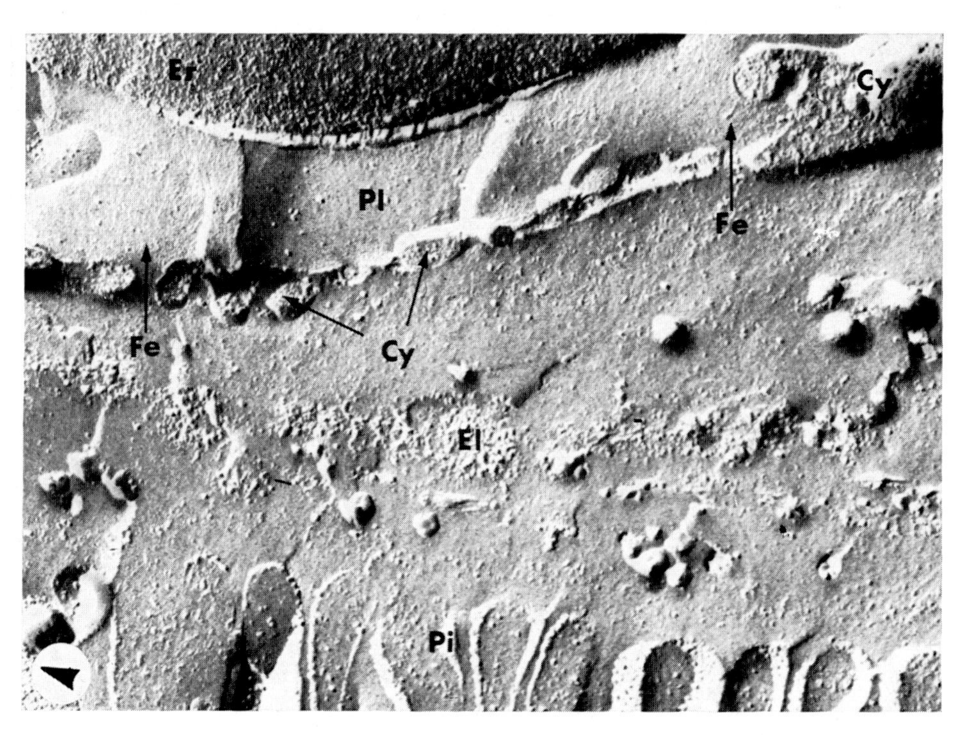

FIG. 89. Replica of capillary in choroid coat of guinea-pig eye. Cy, cytoplasm of endothelial cell; El, lamina elastica choroidea; Er, cross-fractured erythrocyte in capillary lumen; Fe, fenestra; Pi, basal plasma membrane (infolded) of retinal pigment epithelial cell; Pl, blood plasma in lumen of capillary. × 66,500.

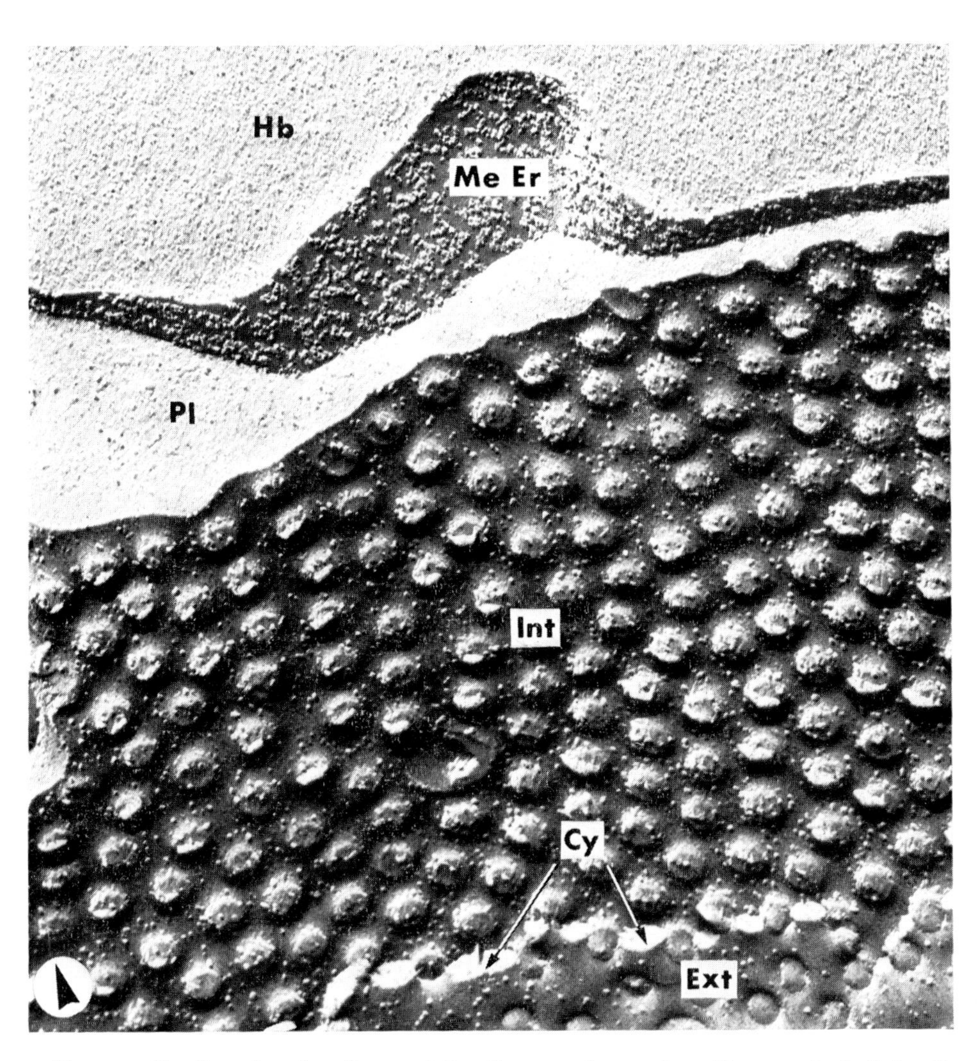

FIG. 90. Replica showing Ext and Int fracture faces of capillary endothelial cell plasma membrane from guinea-pig choroid. Note fenestrae on both faces. Cy, cytoplasm; Hb, haemoglobin of erythrocyte in lumen; Me Er, Ext fracture face of erythrocyte; Pl, blood plasma. × 75,500.

III ONION ROOT (FIGS. 91–94)

Root tips have for a long time provided suitable material for botanists engaged in freeze-fracture studies (Branton and Moor, 1964; Northcote

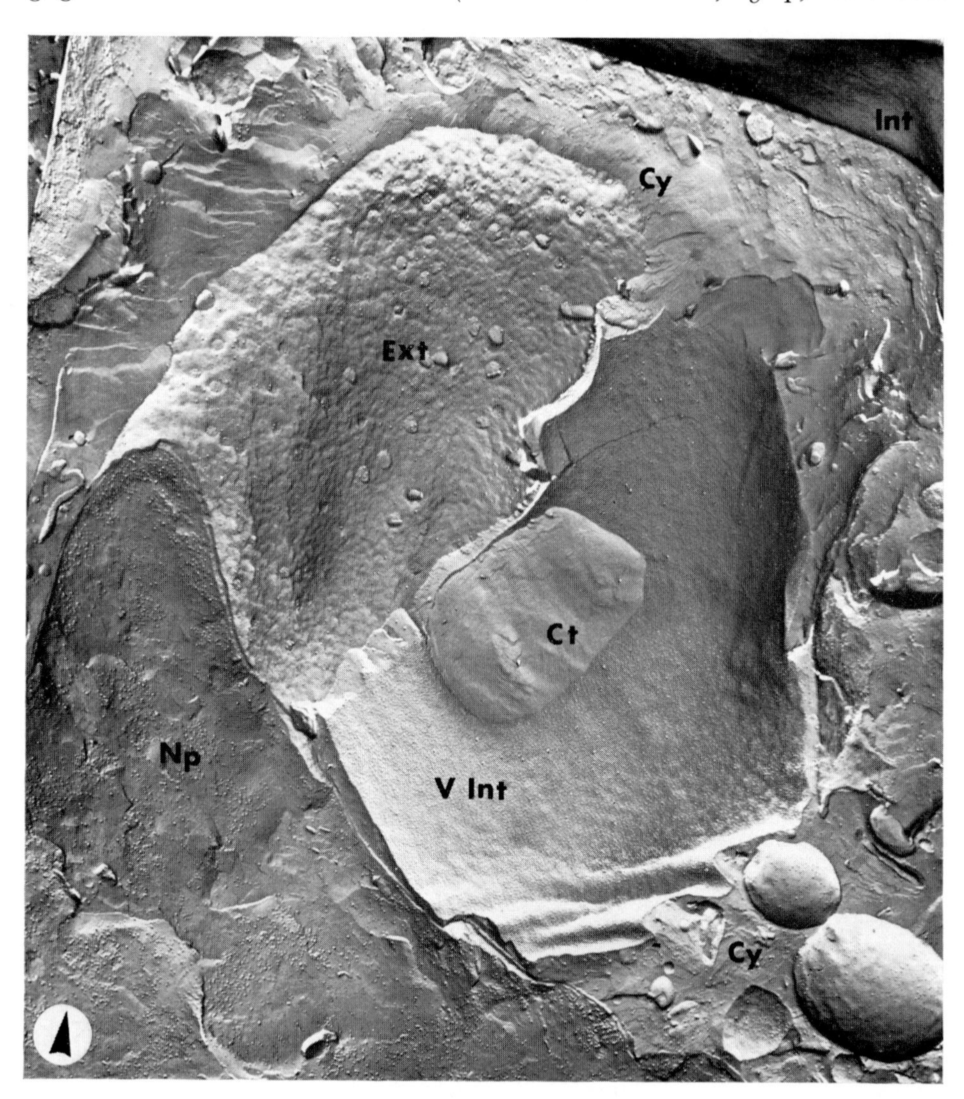

FIG. 91. Cell from root tip of onion. Cy, cytoplasm; Ct, content of cytoplasmic vacuole; Ext, fracture face of nuclear membrane directed towards the exterior and exhibiting pores; Int, fracture face of plasmalemma directed towards the interior; Np, nucleoplasm; V. Int, fracture face of vacuolar membrane (tonoplast) directed towards the interior. × 41,200.

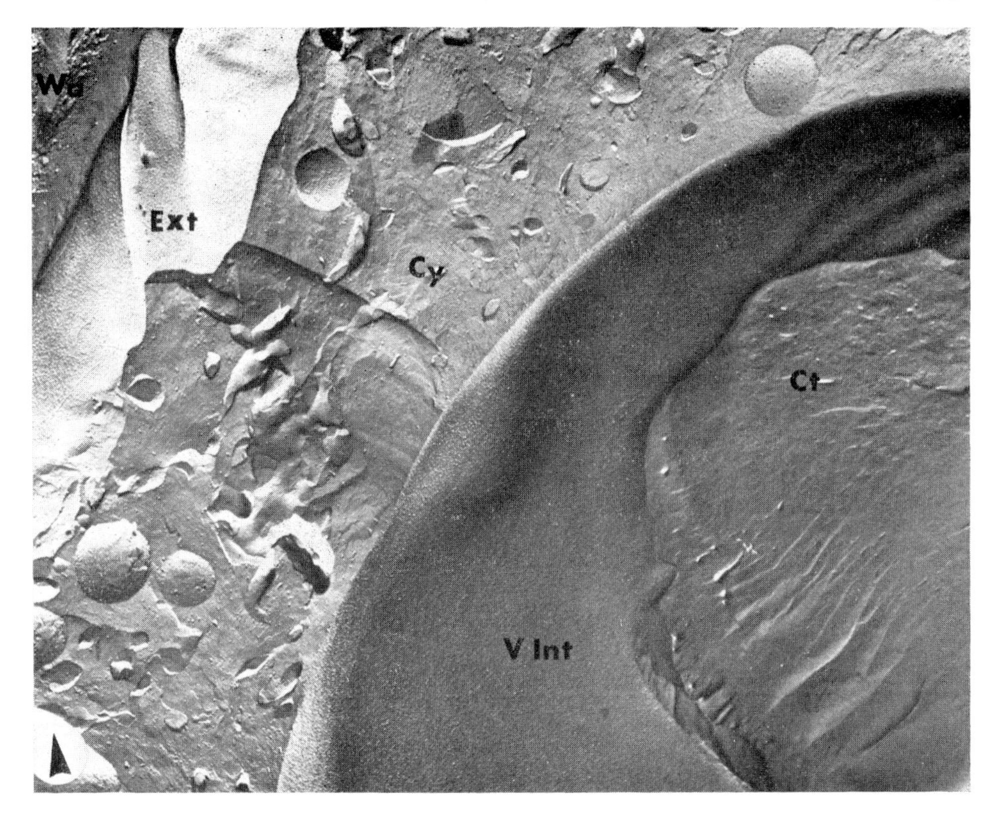

FIG. 92. Cell from root tip of onion. Cy, cytoplasm; Ct, content of cytoplasmic vacuole; Ext, fracture face of plasmalemma directed towards exterior of cell; Wa, cell wall; V. Int, fracture face of vacuolar membrane (tonoplast) directed towards the interior. × 21,000.

and Lewis, 1968; Fineran, 1970, 1973). Nucleus, cytoplasmic organelles, and vacuoles (Figs. 90–93) have provided focal points of interest as has also the plasma membrane. Northcote and Lewis (1968) drew attention to the rows of particles on the 'Ext' face of the latter (Fig. 94) and suggested they may be associated with microfibrillar synthesis.

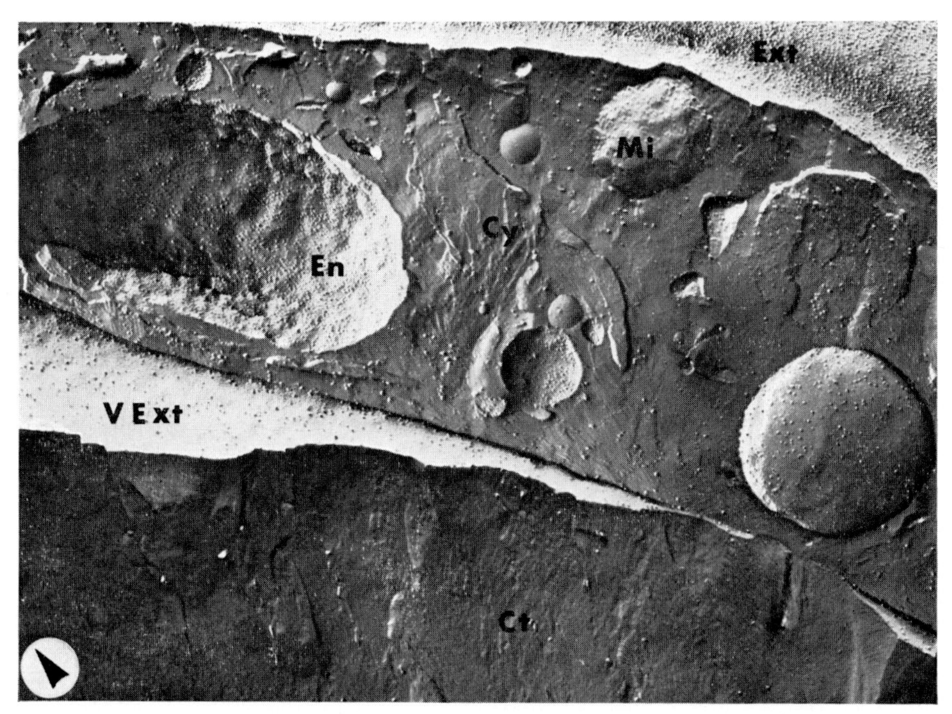

FIG. 93. Cytoplasm of cell of onion root. Cy, cytoplasm; Ct, content of vacuole; En, endoplasmic reticulum; Ext, fracture face of plasmalemma directed towards the exterior; Mi, mitochondrion; V. Ext, fracture face of vacuolar membrane (tonoplast) directed towards the exterior. ×45,000.

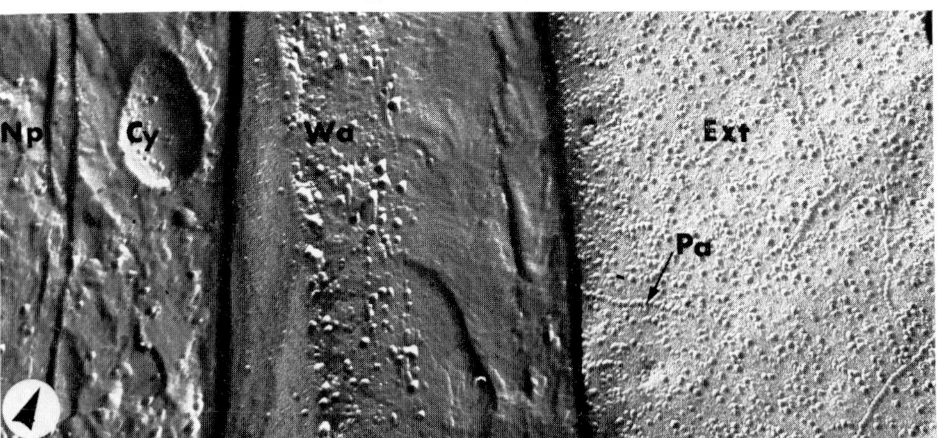

FIG. 94. Replica of onion root cells showing from left to right, nucleoplasm (Np) limited by two leaflets of nuclear membrane; Cy, cytoplasm; Wa, cell walls and fracture face (Ext) of plasmalemma of second cell directed towards the exterior. Note linear rows of particles (Pa) on this fracture face. × 79,000.

IV LIVER (FIGS. 95–97).

Liver is probably the tissue of choice for preparation and examination by the beginner. It is comparatively homogeneous in structure, the cells contain most of the commoner cytoplasmic organelles (Figs. 63, 64 and 95), extensive fracture faces of membranes with specialised

FIG. 95. Replica of mouse liver cell showing general features. En, endoplasmic reticulum; Mi, mitochondrion; Nu, nucleus. × 22,000. For other features of liver see Figs. 53, 63, 64 and 68.

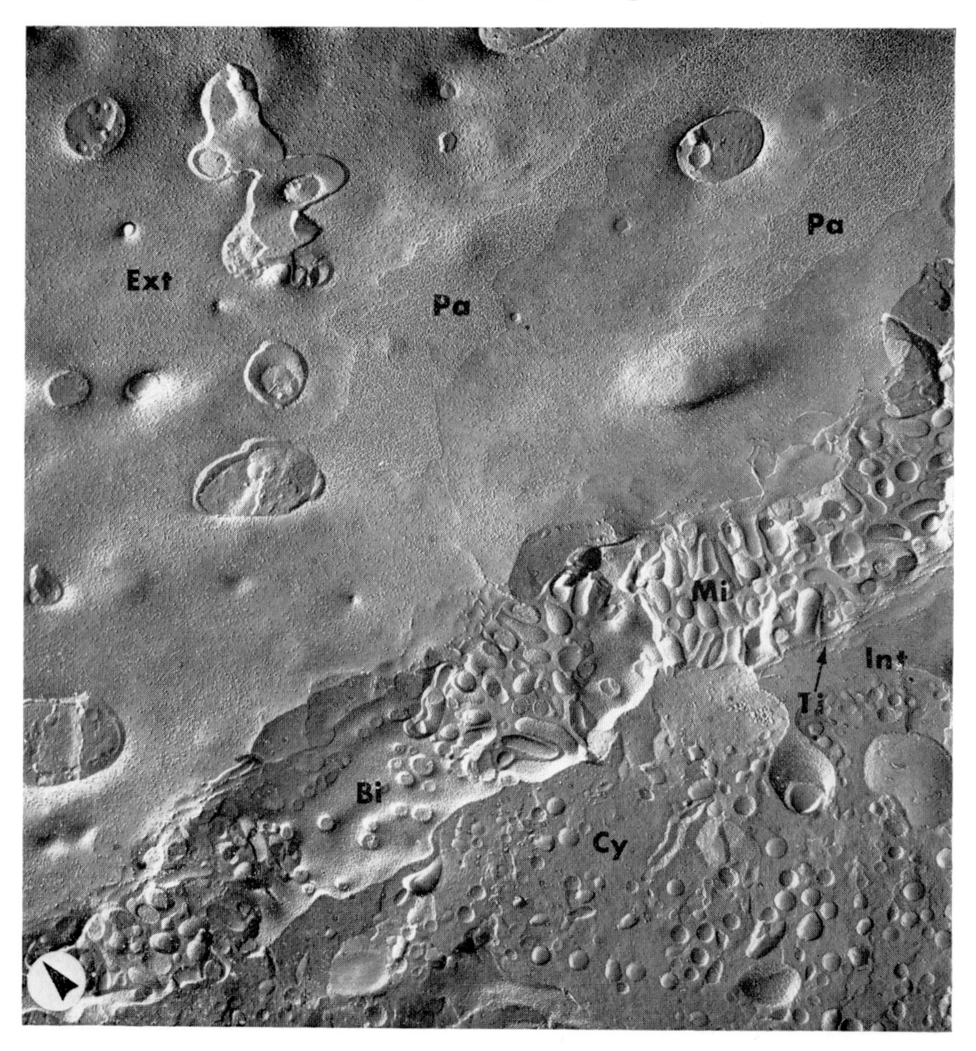

Fig. 96. Replica of mouse liver cells. Bi, bile canaliculus with microvilli projecting into the lumen; Cy, cytoplasm; Ext, Int, appropriately directed fracture faces of plasma membranes of cells; Pa, gap-junction particles on Ext face; Ti, tight junction particles. × 22,000.

contacts are revealed (Figs. 96 and 97) and perhaps most important of all, tissue and replica are quite easily separated so that large pieces of clean replica are readily obtained.

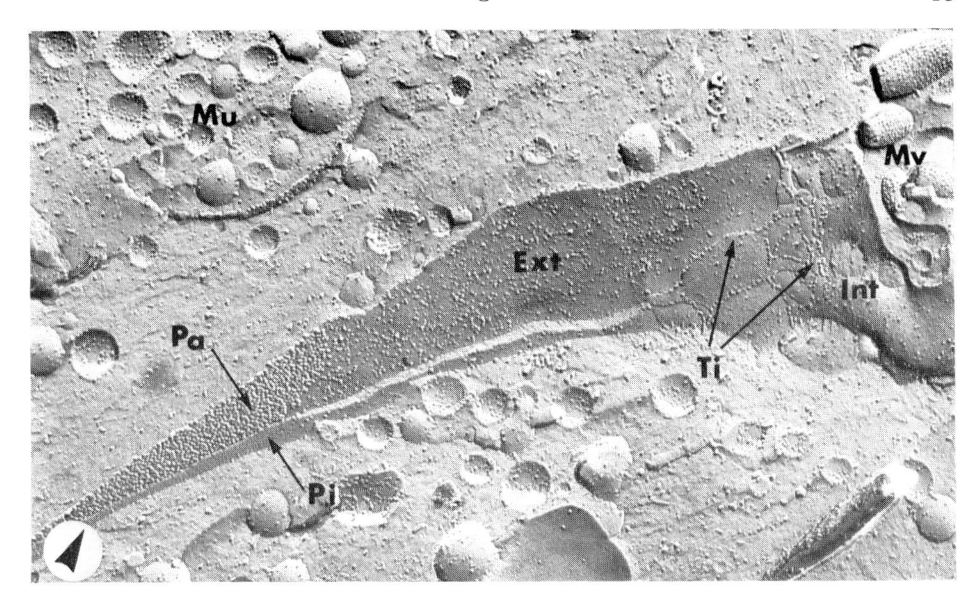

Fig. 97. Apposition of two liver cells in region of bile canaliculus. Ext, Int, appropriately directed fracture faces of apposed membranes; Mu, multivesicular body; Mv, microvilli of canaliculus; Pa, Pi, particles and pits respectively associated with gap junction; Ti, tight junction particles. × 47,500.

V NERVE (FIGS. 98–103)

The basic structural and functional unit of peripheral nerve is the neuron and its satellite cell, the Schwann cell. The two become associated early in development and for a period the relationship between them is comparatively simple, consisting of Schwann cells loosely enveloping bundles of axons (Fig. 98). As development proceeds, this relationship becomes more complicated, particularly where myelinated fibres are concerned. Here, a single axon is invaginated into the Schwann cell (Fig. 99) and the myelin sheath consists of many turns of Schwann plasma membrane wound around the axon.

The laminated character of the myelin sheath is particularly well revealed by the freeze-fracture technique both in peripheral and central nervous systems (Figs. 100 and 102). In the peripheral nervous system, fracture faces of myelin are typically almost completely devoid of particles (Fig. 101) and this is held to be indicative of a low protein content of the membrane (Branton, 1966; Bischoff and Moor, 1967, 1969). In our experience, myelin of the central nervous system exhibits

FIG. 100. Myelinated fibre from peripheral nerve of rabbit. Ax, axon; Cy, Schwann cell cytoplasm; My, myelin sheath. × 86,500.

FIG. 101. Myelin sheath of peripheral nerve of rabbit. Note absence of particles from the fracture faces. × 196,000.

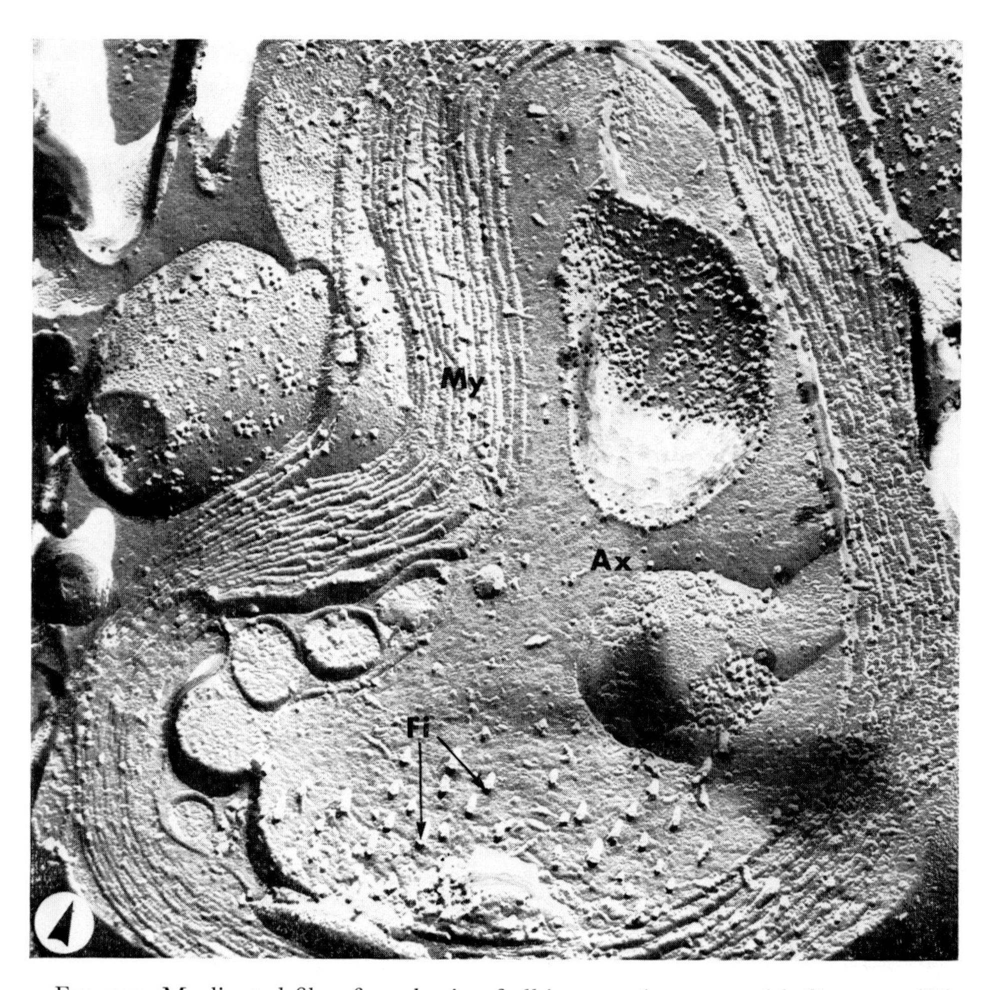

FIG. 102. Myelinated fibre from brain of albino rat. Ax, axon with filaments (Fi) and fracture faces of mitochondria; My, myelin sheath. × 98,000.

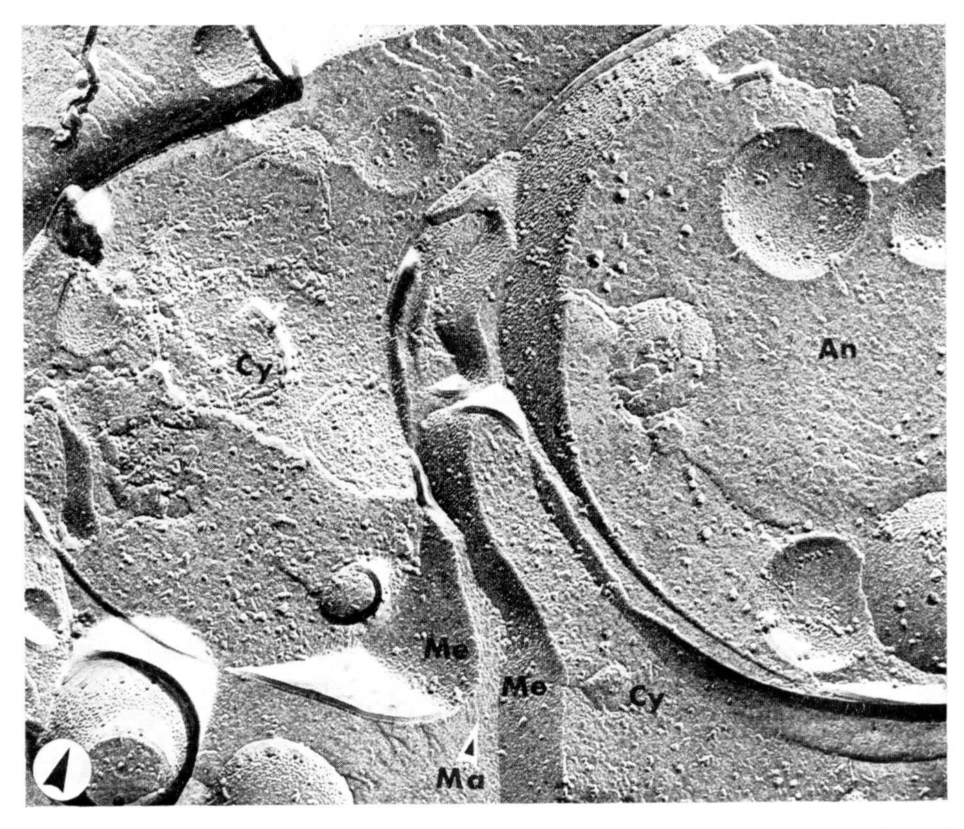

FIG. 103. Unmyelinated axon (An) from human foetal skin. Ma, mesaxon formed by double layer of Schwann cell plasma membrane Me; Cy, Schwann cell cytoplasm. × 112,000.

VI RETINA (FIGS. 104–108)

Multi-layered structures, such as the retina, can present certain difficulties of interpretation, depending upon the plane of fracture, and how many layers happen to be represented on an individual replica. The pigment layer (Fig. 104) is readily recognisable because of its relationship to the underlying choroid coat, and by the presence of characteristically oval-shaped melanosomes in the cytoplasm of the cells. Melanosomes may appear cross-fractured or, 'Ext' and 'Int' faces of the limiting membrane of the organelle may be revealed (Fig. 105). The detailed internal structure seen in stained thin sections is not revealed in replicas (Breathnach *et al.*, 1973b).

The region of the outer segments of the rods and cones is also easily identifiable by virtue of the characteristic stacked lamellae (Fig. 106)

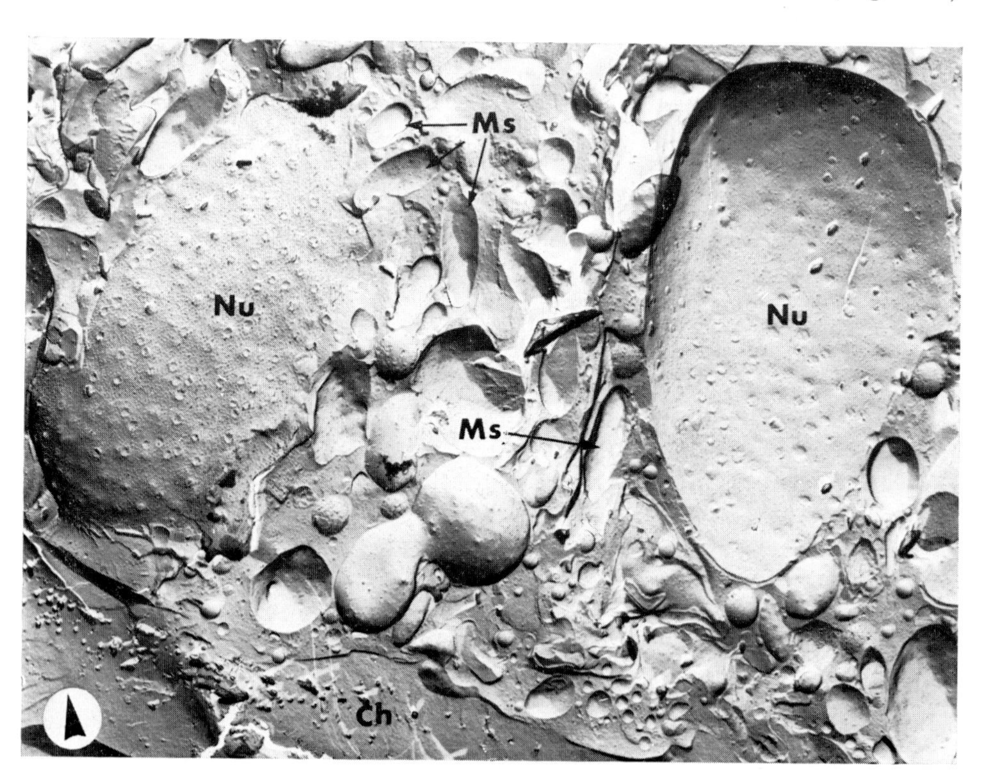

FIG. 104. Replica of basal regions of retinal pigment epithelial cells of human foetus aged 16 weeks. Ch, choroid; Ms, melanosome; Nu, nucleus. × 11,000. Reproduced from Breathnach *et al.* (1973b) *J. Anat.*, with permission.

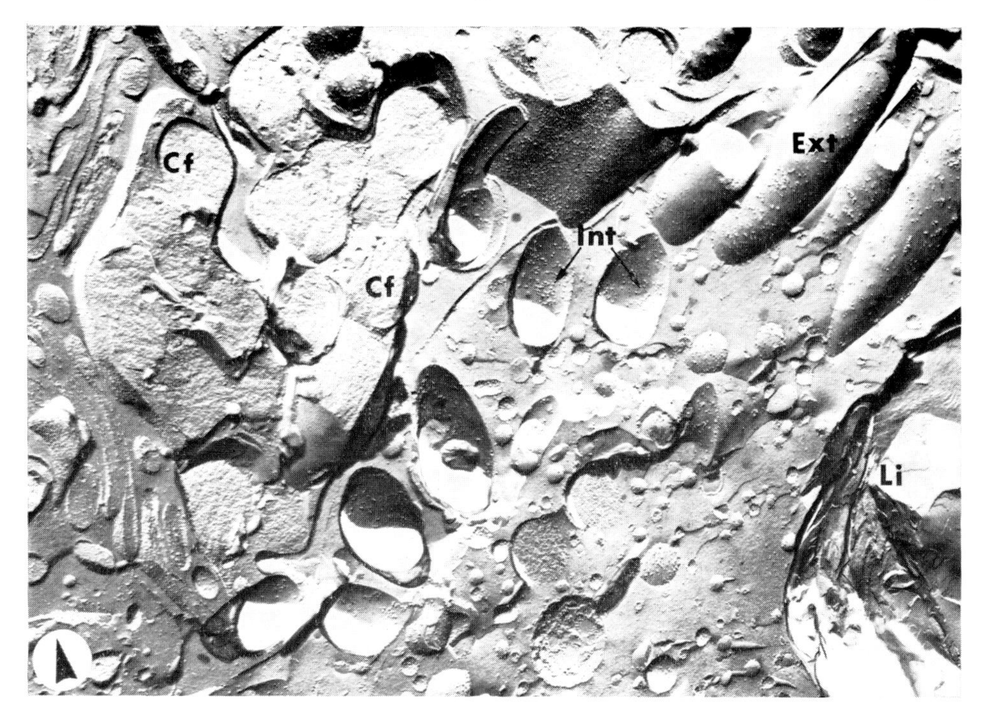

FIG. 105. Replica of apical region of cytoplasm of retinal pigment epithelial cell of mature guinea-pig. Cf, cross-fractured melanosome; Ext, Int, appropriately directed fracture faces of limiting membrane of melanosome; Li, lipid accumulation. × 24,500. Reproduced from Breathnach *et al.* (1973b) *J. Anat.*, with permission.

and the inner segments when cross-fractured are revealed as membrane-limited rounded elements of varying diameter containing many mitochondria (Fig. 107). The outer nuclear layer is most certainly identified from its relation to the rod and cone segments on the one side, and the outer plexiform layer on the other (Figs. 107 and 108). And so on for the other layers.

To date, the neural layers of the retina have only been cursorily studied by freeze-fracture. Preliminary observations indicate that retina provides ideal material for the investigation of synapses by this technique.

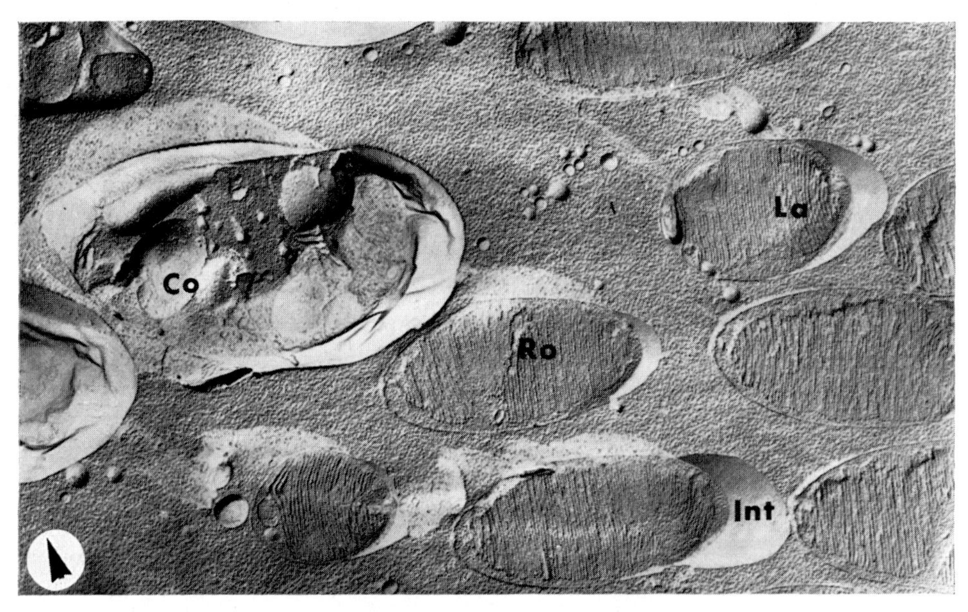

FIG. 106. Replica of rat retina showing outer segments of rods (Ro) and cones (Co). La, laminated membranes; Int, fracture face of limiting membrane of rod, directed towards its interior. × 14,000.

FIG. 107. Replica of guinea-pig retina showing inner segments of rods and cones (In Se) and cells of outer nuclear layer (Ou Nu). × 5300.

FIG. 108. Replica of guinea-pig retina showing cells of outer nuclear layer (Ou Nu) and portion of outer plexiform layer (Ou Pl). × 6000.

VII EPITHELIA

Apart from their individual characteristics, epithelia provide excellent material for the study of specialised cell contacts, and these have been extensively examined by freeze-fracture (see Chapter 4). Here, some general features of a secretory-absorptive (intestinal), and of a stratified squamous (epidermis) epithelium are illustrated.

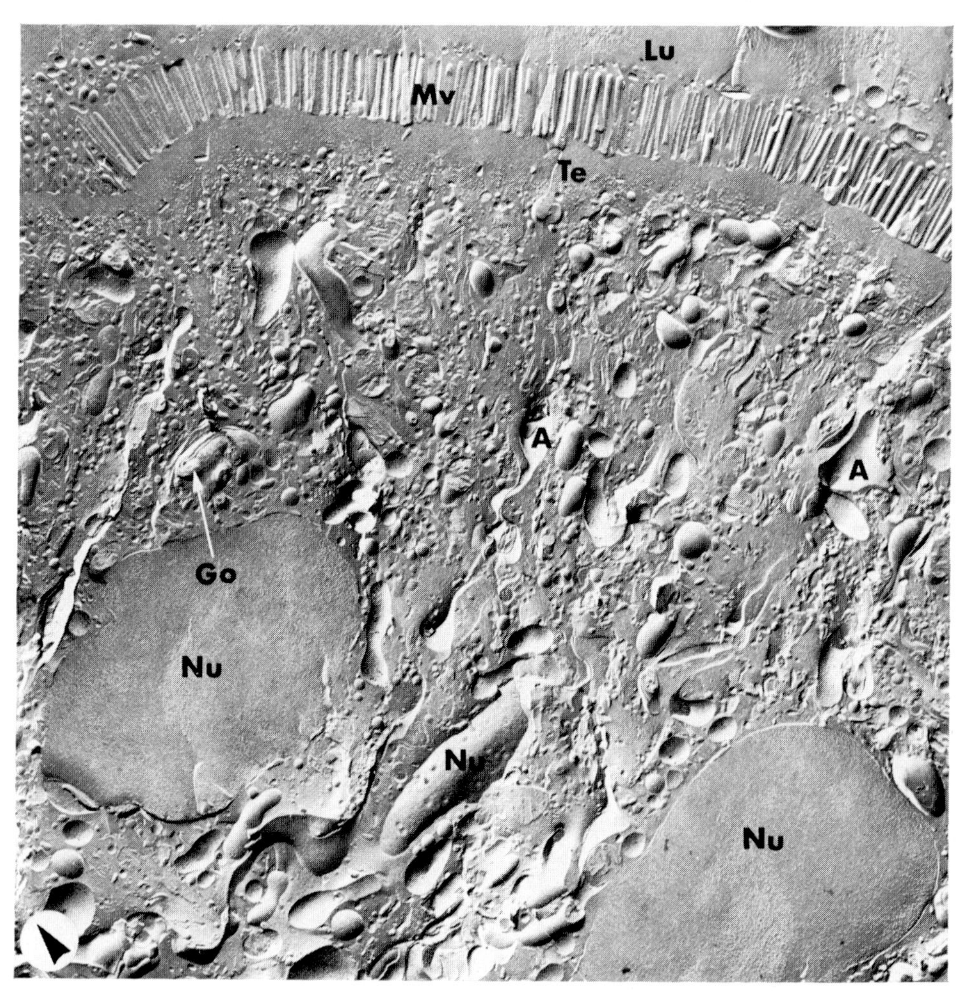

Fig. 109. Replica of jejeunal epithelial cells of mouse. A, indicates general line of cellular appositions; Go, Golgi apparatus; Lu, lumen of intestine; Mo, microvilli of brush border; Nu, nucleus; Te, region described as terminal web in micrographs of thin sections. × 9500. For other features of jejeunal epithelium see Figs. 5, 55 and 69.

A Jejeunal (Figs. 109–111)

Microvilli of the brush border fractured in various planes are particularly well shown by freeze-fracture (Figs. 55 and 111), and an important study of this region has been published by Mukherjee and Staehelin (1971). The cytoplasm of the cells exhibits the common organelles, and Golgi membranes are encountered quite frequently (Figs. 5, 69 and 109). Goblet cells with large mucus droplets or granules are easily distinguished from surrounding cells (Fig. 110).

FIG. 110. Replica of goblet cell (Gob) from mouse jejeunal epithelium. The cytoplasm is full of mucus droplets or granules. Ep, apposed general epithelial cell; Int, fracture face of plasma membrane of goblet cell directed towards the interior. × 16,000.

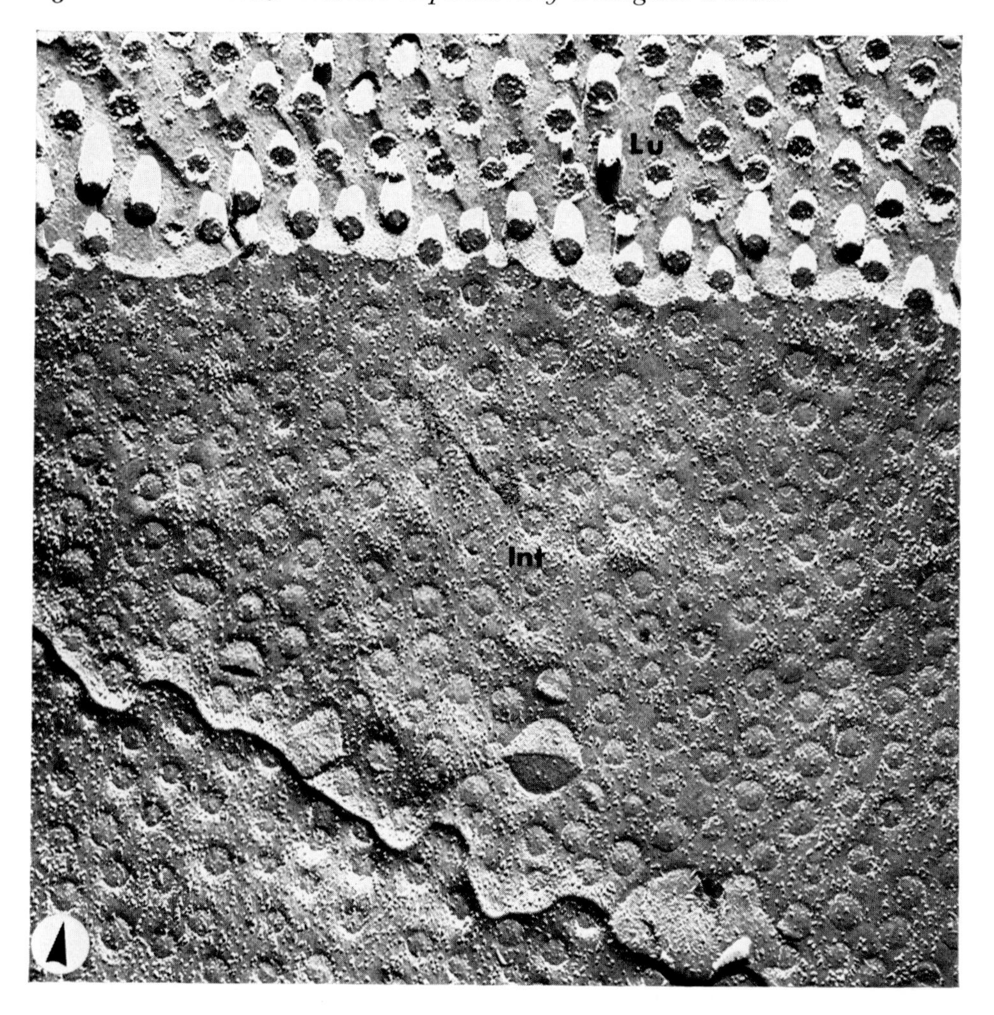

FIG. 111. Replica showing Int fracture face of luminal plasma membrane of jejeunal epithelial cell of mouse. Note regular depressions coinciding with bases of microvilli, tips of which also appear fractured within the lumen, (Lu). × 27,500.

B Epidermis (*Figs. 112–117*)

Epidermis, despite its ready accessibility for biopsy, has only very recently been submitted to examination by freeze-fracture (Reed and Rothwell, 1970; Breathnach; 1973, Breathnach *et al.*, 1972, 1973a). This is mainly due to the fact that skin is not a homogeneous tissue, and certain of its constituent elements, such as collagen and keratin are particularly difficult to dissolve away, or separate from, the replica.

The characteristic cytoplasmic feature of the epidermal keratinocyte is the tono-filament and this is clearly seen in replicas (Figs. 72 and 113). On fracture, tono-filaments are apparently drawn out of the cytoplasm to

FIG. 112. Replica of epidermal keratinocytes of mature human skin. Cy, cytoplasm containing tonofilaments and organelles; Me, fracture faces of apposed plasma membranes; Nu, nucleus. × 19,000. For other features of epidermis see Figs. 49–52 and 72.

some extent before they actually break. They therefore cast prominent and sharp shadows (Fig. 72) and at first sight it would appear that this would allow accurate measurement of their diameters to be made. However, it is probable that they may undergo a degree of plastic deformation during fracturing. Extensive areas of fractured interlocking appositional cell membranes are regularly encountered (Fig.

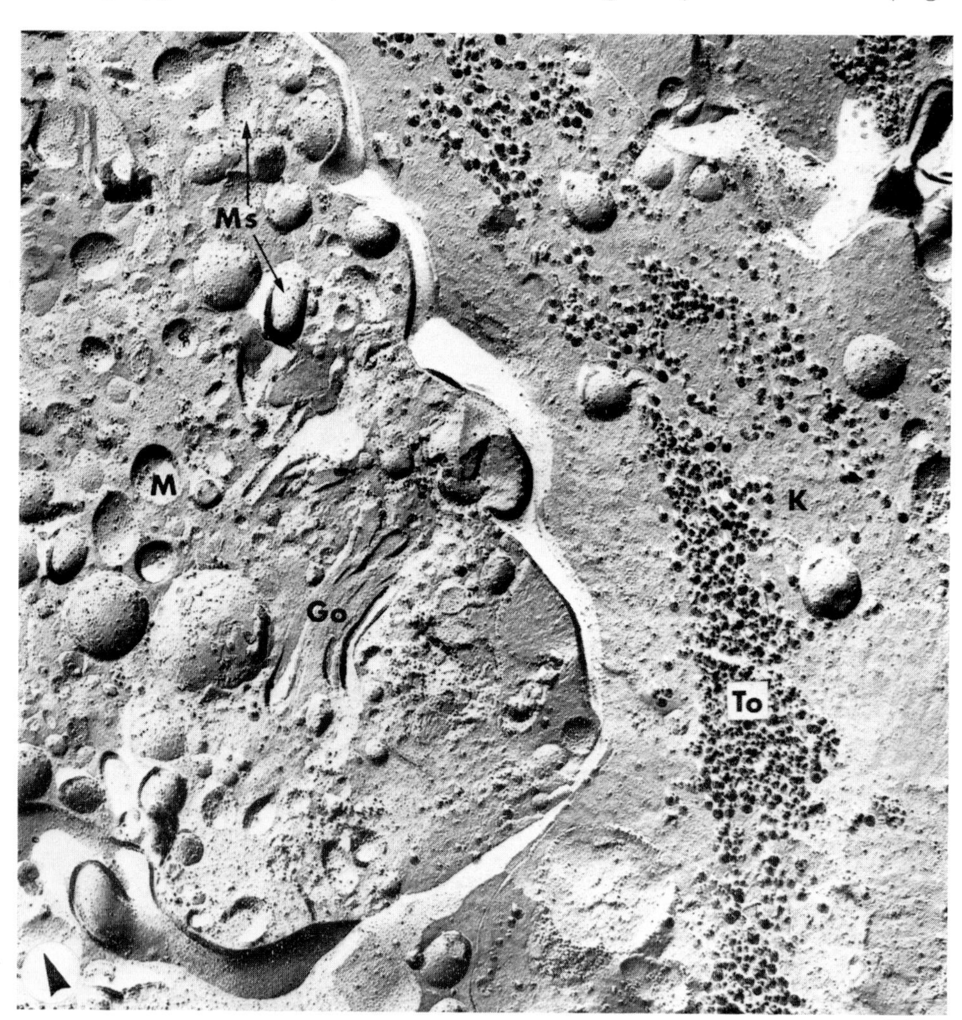

FIG. 113. Replica from epidermis of mature human skin to show differences between keratinocyte (K) and melanocyte (M). Go, Golgi apparatus; Ms, melanosome; To, tonofilaments. × 35,000. Reproduced from Breathnach (1973) *Brit. J. Dermat.*, with permission.

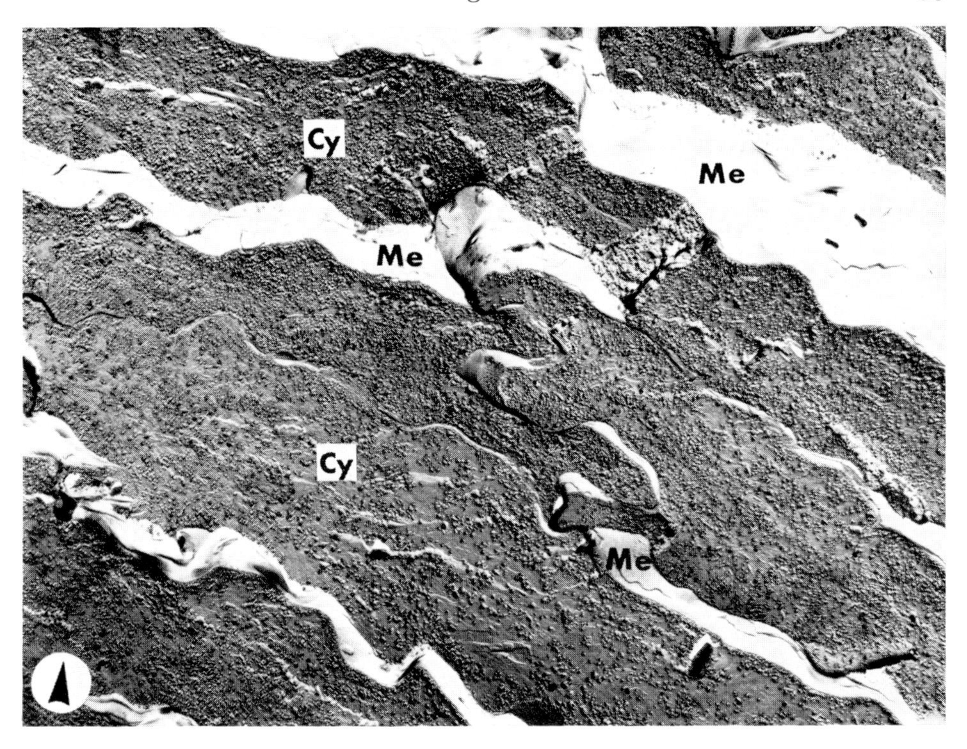

Fig. 114. Replica of stratum corneum of mature human epidermis. Cy, cross-fractured cytoplasm of cells exhibiting filaments; Me, fracture faces of apposed plasma membranes of cells. × 27,000.

112), and on the fracture faces desmosomes and other contacts may be seen (see Chapter 4).

Cells of the stratum corneum may be seen on cross-fracture (Fig. 114) or extensive fracture faces of the plasma membrane may be revealed (Figs. 115 and 116). Virtual absence of M.A.P. from the general plasma membrane, and the presence of desmosomal particles on both 'Ext' and 'Int' faces are characteristic of stratum corneum (Figs. 116 and 117).

Non-keratinocytes, i.e. melanocytes (Fig. 113) and Langerhans cells (Reed and Rothwell, 1970; Caputo, 1975) are readily distinguishable by virtue of an absence of tono-filaments and the presence, in each instance of a specific identifying cytoplasmic organelle.

Fig. 115. Replica of stratum corneum of mature human epidermis. In this replica considerably larger areas of membrane fracture face (Me) are exposed. Cy, cytoplasm. × 27,000. Reproduced from Breathnach *et al.* (1973a) *J. Anat.*, with permission.

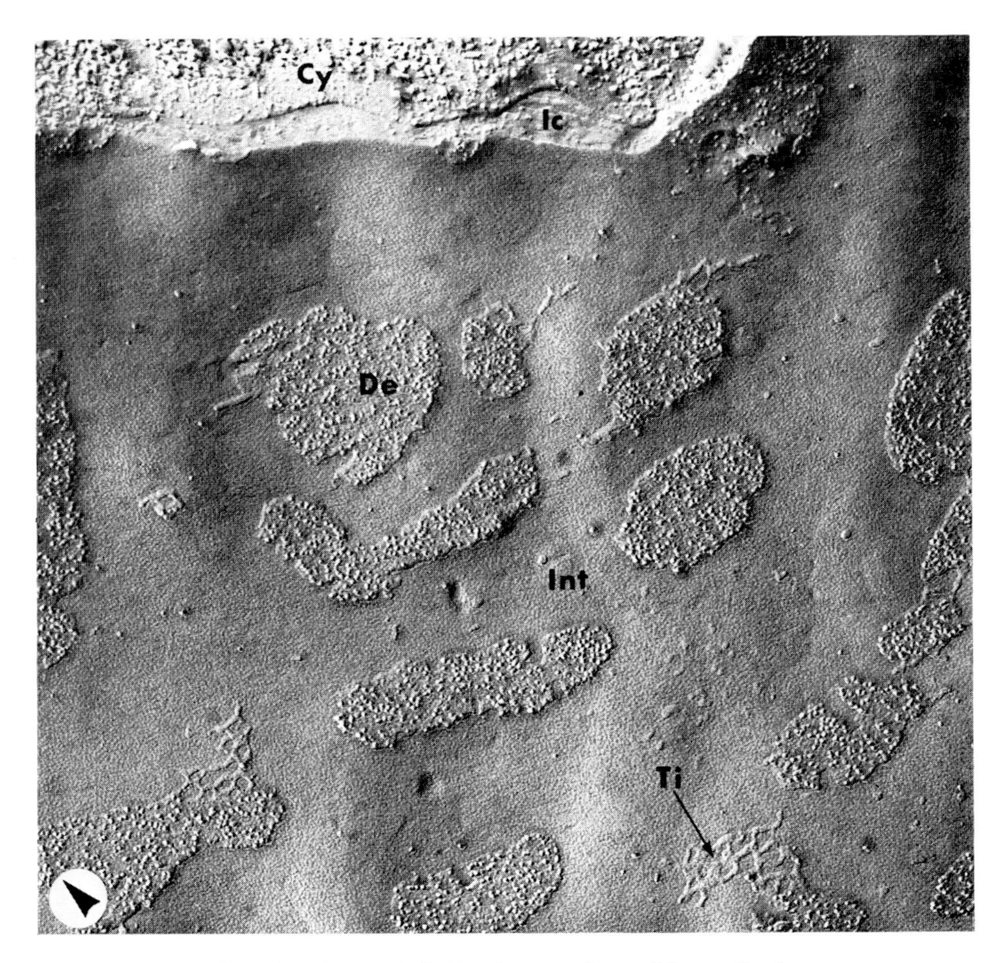

FIG. 116. Replica showing mainly Int fracture face of first cell of stratum corneum of mature human epidermis. Note absence of individual M.A.P. from the general fracture face. Cy, cytoplasm of uppermost stratum granulosum cell; De, aggregated desmosomal particles; Ic, intercellular material; Ti, possibly modified tight junction. × 65,000.

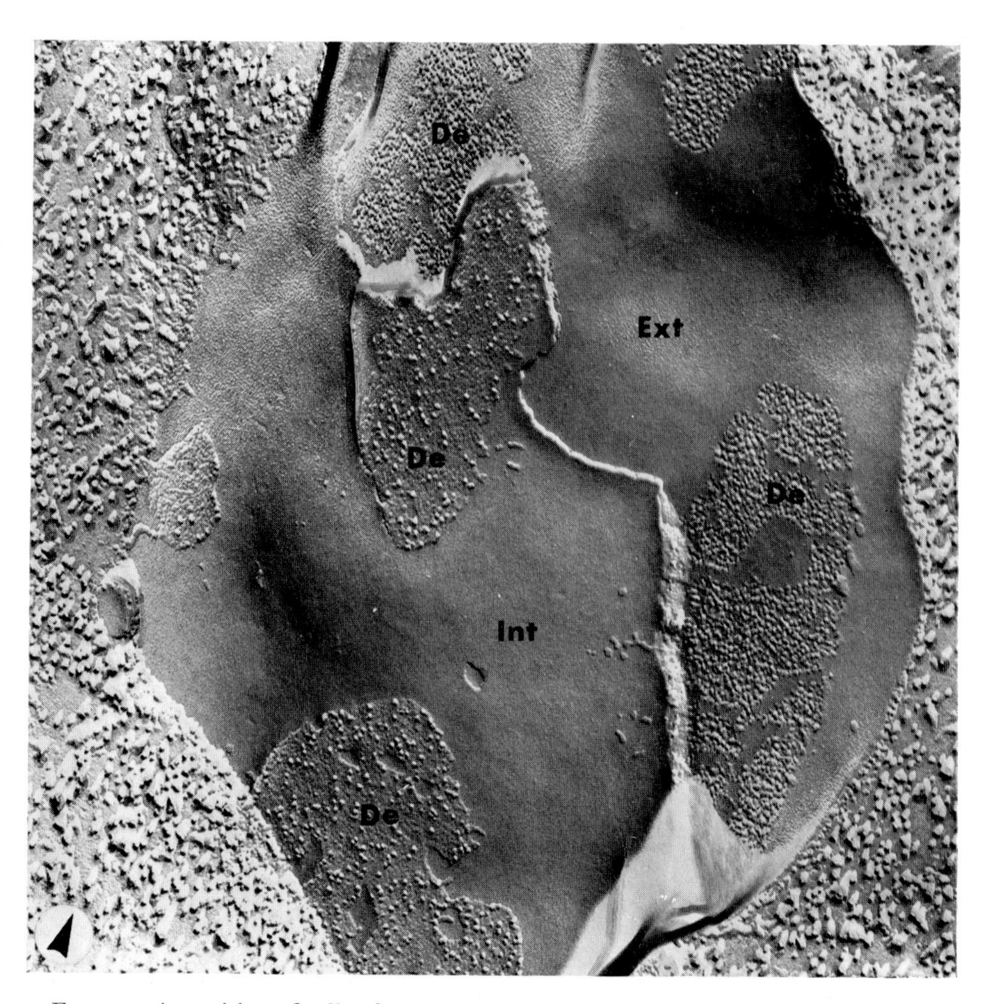

FIG. 117. Apposition of cells of stratum corneum of mature human epidermis. Note presence of desmosomal particles (De) on both (Ext, Int) fracture faces of the plasma membrane. × 75,000. Reproduced from Breathnach *et al.* (1973b) *J. Anat.*, with permission.

VIII CONNECTIVE TISSUE (FIGS. 118–121)

Connective tissue is made up of collagen fibrils and elastic fibres, with interspersed fibroblasts. Collagen fibrils may appear cross-fractured or longitudinally fractured (Figs. 118–120), and in the latter instance the overall appearance of a bundle is ragged due to the fact that many fibrils are pulled out of the matrix on fracturing; they and their long shadows overlap and obscure fibrils lying flush with the general fracture plane. Freeze-fracture has established that there is a helical component in the substructure of collagen fibrils (Szirmai *et al.*, 1970; Breathnach *et al.*, 1972; Raynes, 1974), and this is evidently due to an arrangement of filaments (Figs. 119 and 120). A similar helical arrangement has been demonstrated by Bouteille and Pease (1971) in thin sections of material prepared by an inert dehydration technique. Cross-banding of fibrils is rarely seen in replicas. We have found it more

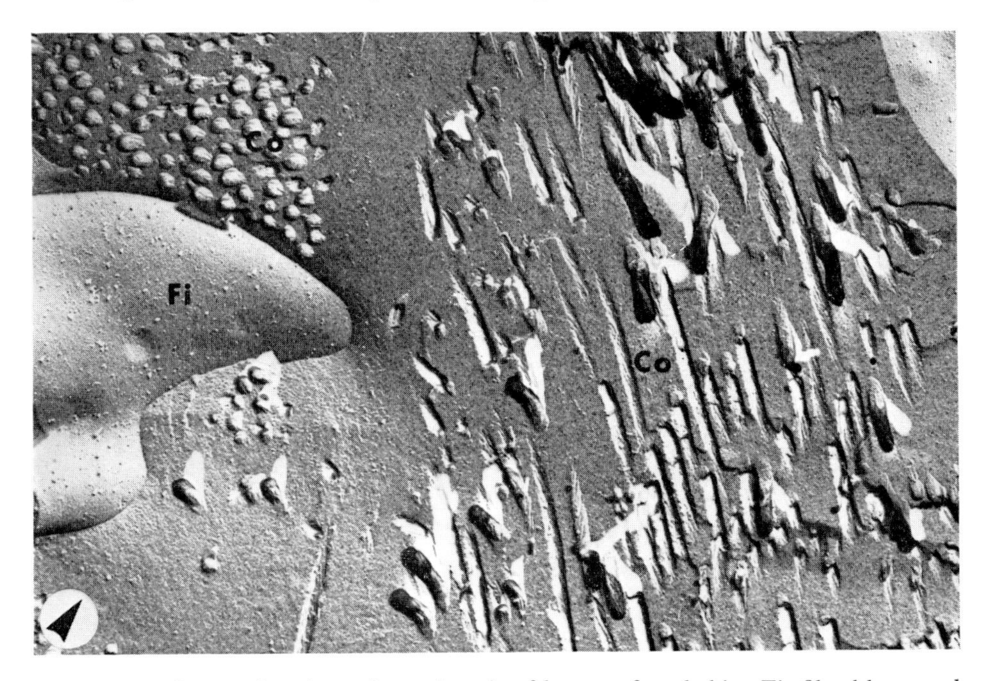

FIG. 118. Connective tissue from dermis of human foetal skin. Fi, fibroblast, and above it collagen (Co) cross-fractured. On the right side of the field, are collagen fibrils fractured longitudinally (or their imprints) and the helical arrangement is apparent. × 43,000.

Fig. 119. Cross-fractured collagen fibrils from mature human skin. Note apparent inner (In) and outer (Ou) components of some fibrils. Fi, filaments. × 163,000

frequently in sublimated material whereas Raynes (1974) reports the contrary. Cross-fractured collagen fibrils have an irregular appearance, and sometimes give the impression of being composed of inner and outer components (Fig. 119).

Elastic fibres (Fig. 121) appear in replicas as aggregates of closely packed filaments.

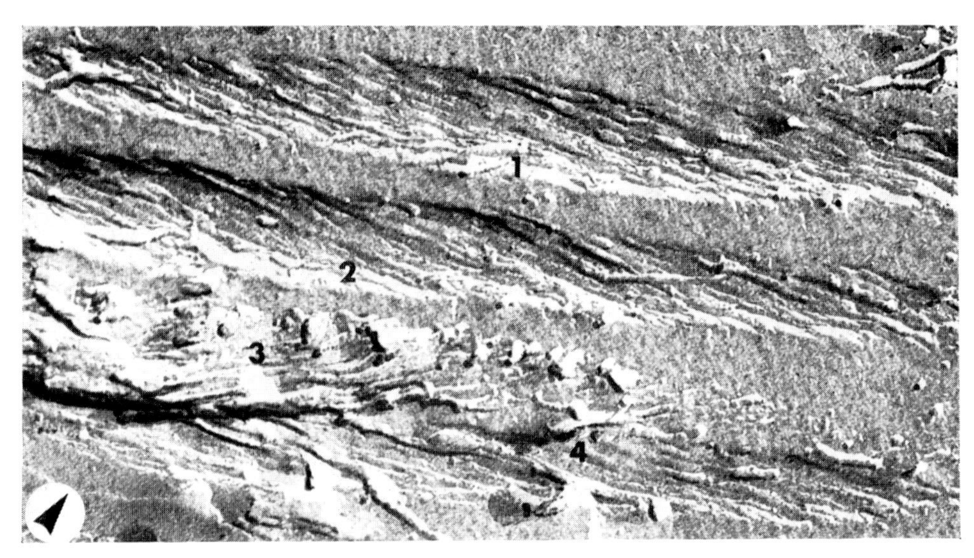

FIG. 120. Longitudinally fractured collagen from mature human skin. Fibrils 1, 2 and 4 have been fractured away leaving imprints. Fibril 3 is present. In either case the helical arrangement due to filaments is apparent, particularly if viewed sideways × 64,500.

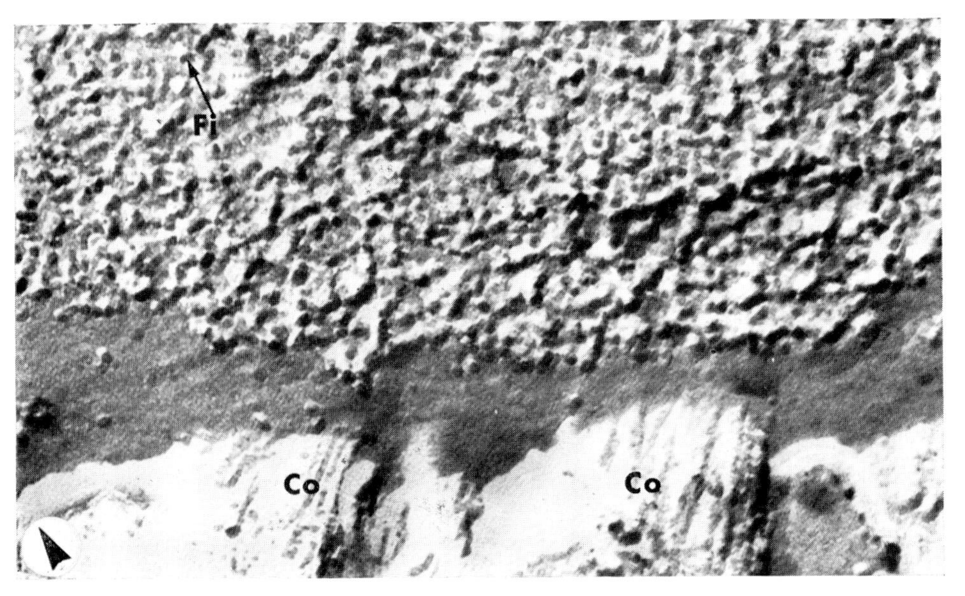

FIG. 121. Cross-fractured elastic fibre from mature human skin. It is made up of closely packed filaments (Fi). Co, collagen fibrils with evident filamentous substructure. × 241,000.

REFERENCES

Bischoff, A. and Moor, H. (1967). *Z. Zellforsch.* **81**, 303–310.
Bischoff, A. and Moor, H. (1969). *Med. Biol. Illust.* **19**, 89–94.
Bouteille, M. and Pease, D. C. (1971). *J. Ultrastruc. Res.* **35**, 314–338.
Branton, D. (1966). *Exptl. Cell Res.* **45**, 703–707.
Branton, D. (1971). *Phil. Trans. Roy. Soc. Lond.* **B261**, 133–138.
Branton, D. and Moor, H. (1964). *J. Ultrastruct. Res.* **11**, 401–411.
Breathnach, A. S. (1973). *Brit. J. Dermat.* **88**, 563–574.
Breathnach, A. S., Goodman, T., Stolinski, C. and Gross, M. (1973a). *J. Anat.* **114**, 65–81.
Breathnach, A. S., Gross, M. and Martin, B. (1973b). *J. Anat.* **116**, 303–320.
Breathnach, A. S., Stolinski, C. and Gross, M. (1972) *Micron* **3**, 287–304.
Caputo, R. (1975). Personal communication.
Fineran, B. A. (1970). *J. Ultrastruct. Res.* **33**, 574–586.
Fineran, B. A. (1973). *J. Ultrastruct. Res.* **43**, 75–87.
Giesbrecht, P. (1966). Proc. 6th Int. Cong. Elect. Micr. Kyoto **2**, 341.
Hoak, C. J. (1972). *Blood* **10**, 514–522.
Huhn, D. and Grassmann, D. (1969). *Blut* **18**, 211–217.
Kirk, R. G. and Ginzberg, M. (1972). *J. Ultrastruct. Res.* **41**, 80–94.
Lickfeld, K. G., Achterrath, M., Hentrich, F., Kolehmainen-Seveus, L. and Persson, J. (1972). *J. Ultrastruct. Res.* **38**, 27–45.
Moor, H. and Mühlethaler, K. (1963). *J. Cell. Biol.* **17**, 609–628.
Mukherjee, T. M. and Staehelin, L. A. (1971). *J. Cell Sci.* **8**, 573–599.
Nanninga, N. (1971). *J. Cell Biol.* **49**, 564–570.
Nanninga, N. (1973). *In* 'Freeze-etching, techniques and applications' (E. L. Benedetti and P. Favard, eds) 151–179. Société Française de Microscopie Électronique, Paris.
Northcote, D. N. and Lewis, P. R. (1968). *J. Cell Sci.* **3**, 199–206.
Pinto da Silva, P. and Branton, D. (1970). *J. Cell Biol.* **45**, 598–605.
Raynes, D. G. (1974). *J. Ultrastruct. Res.* **48**, 59–66.
Reed, R. and Rothwell, P. J. (1970). *Brit. J. Dermat.* **82**, 470–486.
Remsen, C. C. and Lundgren, D. G. (1966). *J. Bacteriol.* **92**, 1765.
Remsen, C. C. and Watson, S. W. (1973). *Int. Rev. Cytol.* **33**, 253–296.
Sandri, C., Akert, K., Livingstone, R. B. and Moor, H. (1972). *Brain Research* **41**, 1–16.
Simionescu, M., Simionescu, N., Palade, G. E. (1974). *J. Cell Biol.* **60**, 128–52.
Smith, U., Ryan, J. W. and Smith, D. S. (1973). *J. Cell Biol.* **56**, 492–499.
Szirmai, J. A., van Raamsdonk, W. and Golavarzi, G. (1970). *J. Cell Biol.* **47**, abstract 209a.
Tillack, T. W. and Marchesi, V. T. (1970). *J. Cell Biol.* **45**, 649–653.
Watson, S. W. and Remsen, C. C. (1970). *J. Ultrastruct. Res.* **33**, 148–160.
Weinstein, R. S. and Bullivant, S. (1967). *Blood* **29**, 780–789.

Appendix

A PHYSICAL CONSTANTS OF FREEZING FLUIDS

Fluid	*Formula*	*Mol. Wt*	*Melt. Pt*	*Boil. Pt*
Oxygen*	O_2	32·0	−218·4 °C	−183·0 °C*
Nitrogen	N_2	28·0	−210·0	−195·9
Propane	C_3H_8	44·1	−189·9	−42
Freon 14	CF_4	88·0	−184·2	−128·0
Freon 13	$CClF_3$	104·4	−181·2	−81·4
Freon 22	$CHClF_2$	86·5	−160·0	−40·0
2-Methylbutane (Isopentane)	C_5H_{12}	72·2	−160·0	+27 to +30
Freon 12	CCl_2F_2	120·9	−158·2	−29·8

*When liquifying on other cold fluid surfaces, Oxygen may form a dangerous explosive mixture.

B MELTING AND BOILING POINTS OF EVAPORANTS

Substance	*Melting Pt °C*	*Boiling Pt °C*
Carbon	3527	4220
Tungsten	3370	5900
Tantalum	2996	4100
Iridium	2454	4800
Platinum	1773	4300

C ICE RECESSION IN VACUUM

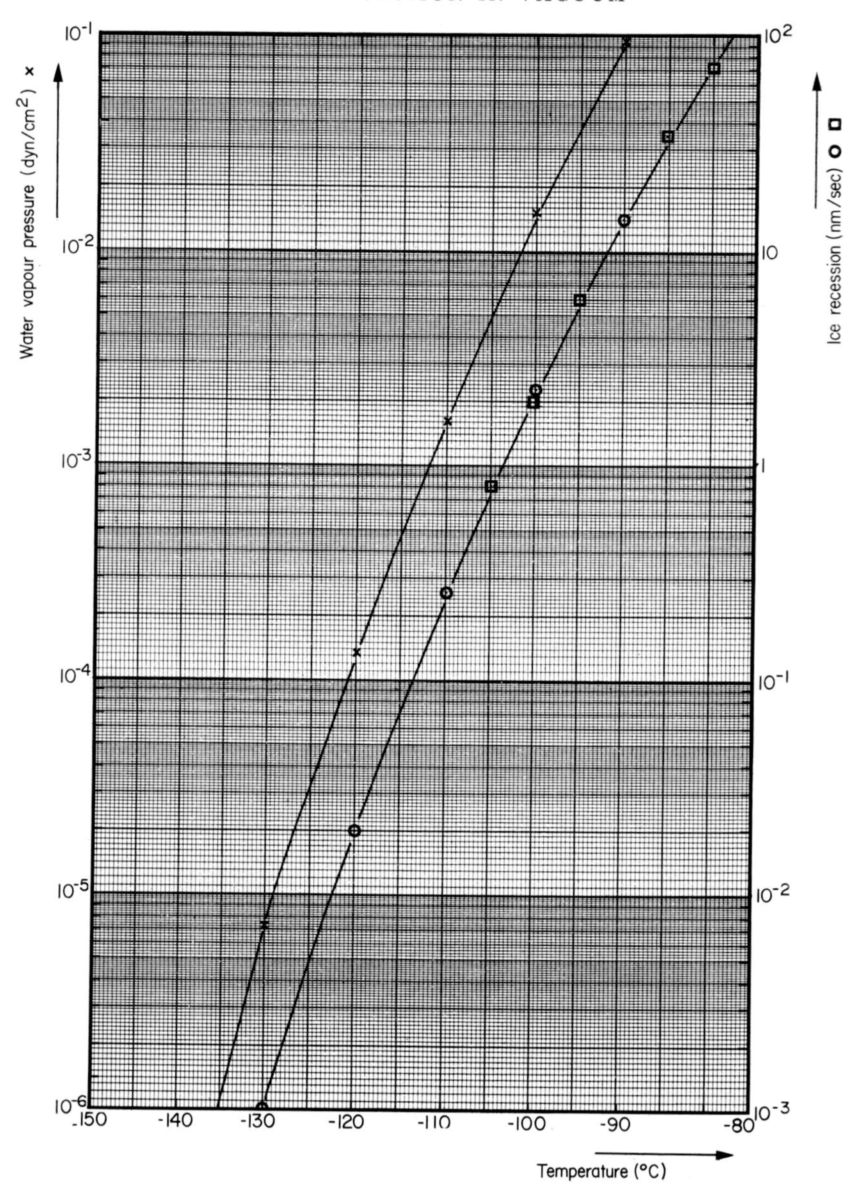

The graph above illustrates rate of recession of ice surface in vacuum and change of water vapour pressure with temperature in the range −140°C to −80°C. ○, calibration according to Hall (1950); □, calibration according to Davy and Branton (1970).

Index